T0239388

Intelligent Robot

Feng Duan • Wenyu Li • Ying Tan

Intelligent Robot

Implementation and Applications

 Springer

Feng Duan
College of Artificial Intelligence
Nankai University
Tianjin, China

Wenyu Li
College of Artificial Intelligence
Nankai University
Tianjin, China

Ying Tan
College of Artificial Intelligence
Nankai University
Tianjin, China

ISBN 978-981-19-8255-2   ISBN 978-981-19-8253-8   (eBook)
https://doi.org/10.1007/978-981-19-8253-8

Jointly published with Beijing Huazhang Graphics & Information Co., Ltd, China Machine Press
The print edition is not for sale in China (Mainland). Customers from China (Mainland) please order the
print book from: Beijing Huazhang Graphics & Information Co., Ltd, China Machine Press.

© China Machine Press, Beijing, China 2023, corrected publication 2023
This work is subject to copyright. All rights are solely and exclusively licensed by the Publisher, whether
the whole or part of the material is concerned, specifically the rights of reprinting, reuse of illustrations,
recitation, broadcasting, reproduction on microfilms or in any other physical way, and transmission
or information storage and retrieval, electronic adaptation, computer software, or by similar or
dissimilar methodology now known or hereafter developed.
The use of general descriptive names, registered names, trademarks, service marks, etc. in this publication
does not imply, even in the absence of a specific statement, that such names are exempt from the relevant
protective laws and regulations and therefore free for general use.
The publishers, the authors, and the editors are safe to assume that the advice and information in this
book are believed to be true and accurate at the date of publication. Neither the publishers nor the
authors or the editors give a warranty, expressed or implied, with respect to the material contained
herein or for any errors or omissions that may have been made. The publishers remain neutral with
regard to jurisdictional claims in published maps and institutional affiliations.

This Springer imprint is published by the registered company Springer Nature Singapore Pte Ltd.
The registered company address is: 152 Beach Road, #21-01/04 Gateway East, Singapore 189721,
Singapore

Preface

Robotics has developed rapidly since the mid-twentieth century when research on modern robots was conducted. Nowadays, robots are integrated into our work and daily life. With the development of computer, Internet, and artificial intelligence technologies, there are more and more types of robots with enhanced functions and an improved experience of using them.

At the same time, new advances in robotics research and development continue to be made. In particular, the Robot Operating System (ROS) is of great strategic importance to the development of the intelligent robotics industry. Robot operating system can provide a unified platform for robot development, so that more users can conveniently study and verify robot algorithms and develop robot applications on this platform, which greatly promotes the development of robot technology. In 2010, Willow Garage released the open-source robot operating system ROS (Robot Operating System). Rather than being an operating system, ROS is a distributed, modular open-source software framework. With the advantages of peer-to-peer design, programming language independence, and open source, ROS has become a new hot spot for learning and use in robotics research.

The Laboratory of Intelligent Perception and Human–Robot Interaction, College of Artificial Intelligence, Nankai University, where the authors' work is dedicated to the research of intelligent robots. This book is based on our long-term experience in developing robots using ROS, and we hope to provide a useful reference book for university students who are interested in learning intelligent robotics and technical staff who are engaged in intelligent robot development work.

This book is divided into three parts: the first part mainly introduces the basics of robotics, including the definition of robotics, development history, key technologies, framework and use of ROS, etc.; the second part introduces how to develop a robot with relatively complete functions in terms of robot hardware and software composition, vision function implementation, autonomous navigation function implementation, voice interaction function implementation, grasping function implementation, etc.; the third part gives comprehensive cases with different application scenarios to show how to develop robots with different functions.

The cases in this book are based on the robot program that the author's team participated in and won the RoboCup Robotics World Open. Beginners following the explanations in this book and practicing with the hands-on resources that accompany this book can not only master the advanced concepts of the software framework involved in robot development but also develop a fully functional intelligent robot step by step.

Tianjin, China   Feng Duan
                                                                       Wenyu Li
                                                                        Ying Tan

---

The original version of this book has been revised with book copyright holder has been updated. An correction to this book can be found at https://doi.org/10.1007/978-981-19-8253-8_15

Contents

Part I   Basic Knowledge

1   Overview of Robots   3
   1.1   Definition and Classification of Robots   3
       1.1.1   Definition of Robots   3
       1.1.2   Classification of Service Robots   4
   1.2   The Evolution of Modern Robots   5
       1.2.1   Early Stages of Modern Robots Research   5
       1.2.2   The 1970s   6
       1.2.3   The 1980s   7
       1.2.4   The 1990s   11
       1.2.5   Twenty-First Century   15
   1.3   Composition of the Robots   23
       1.3.1   Actuators for the Robot   24
       1.3.2   Drive Units   24
       1.3.3   Sensing Devices   24
       1.3.4   Control Systems   25
       1.3.5   Intelligent Systems   25
       1.3.6   Intelligent Human-Machine Interface Systems   26
   1.4   Key Technologies for Robots   26
       1.4.1   Autonomous Mobile Technologies   26
       1.4.2   Perceptual Technologies   27
       1.4.3   Intelligent Decision-Making and Control
             Technologies   27
       1.4.4   Communication Technologies   27
   1.5   Trends in Robots   28
       1.5.1   Hierarchical and Humanized Human-Computer
             Interaction   28

	1.5.2	Interaction Intelligence with the Environment	28
	1.5.3	Networking of Resource Utilization	28
	1.5.4	Standardization, Modularization and Systematization of Design and Production	29
	Further Reading		29

2 **Getting Started with ROS** 31

2.1 Introduction to ROS 32
- 2.1.1 Why Use ROS 32
- 2.1.2 What Is ROS 32
- 2.1.3 Differences Between ROS and Computer Operating Systems 33
- 2.1.4 Key Features of ROS 33

2.2 Installation and Uninstallation of ROS 35
- 2.2.1 Versions of ROS 35
- 2.2.2 Installing and Configuring ROS Indigo 35
- 2.2.3 Installing and Configuring ROS Melodic 38
- 2.2.4 Uninstallation of ROS 40

2.3 Resources for Further Learning 40

Further Reading 41

3 **The Framework and Fundamental Use of ROS** 43

3.1 ROS Framework 43
- 3.1.1 File System Level 43
- 3.1.2 Calculation of Graph Levels 44
- 3.1.3 Community Level 46

3.2 Basics of ROS Use 47
- 3.2.1 Overview of Catkin 47
- 3.2.2 Workspaces and Their Creation Methods 47
- 3.2.3 Creating ROS Project Packages 50
- 3.2.4 Compiling the ROS Project Package 51
- 3.2.5 Creating ROS Nodes 51
- 3.2.6 Compiling and Running ROS Nodes 52
- 3.2.7 Use of Roslaunch 53
- 3.2.8 Creating ROS Messages and Services 55
- 3.2.9 Writing a Simple Message Publisher and Subscriber (C++ Implementation) 59
- 3.2.10 Writing a Simple Message Publisher and Subscriber (Python Implementation) 61
- 3.2.11 Testing Message Publisher and Subscriber 63
- 3.2.12 Writing a Simple Server and Client (C++ implementation) 64
- 3.2.13 Writing a Simple Server and Client (Python Implementation) 66
- 3.2.14 Testing a Simple Server and Client 68

Further Reading 69

4   ROS Debugging   71
   4.1   Common Commands for ROS Debugging   71
   4.2   Common Tools for ROS Debugging   73
          4.2.1   Using rqt_console to Modify the Debug Level
                   at Runtime   73
          4.2.2   Using roswtf to Detect Potential Problems in the
                   Configuration   76
          4.2.3   Displaying Node State Graphs Using rqt_graph   76
          4.2.4   Plotting Scalar Data Using rqt_plot   78
          4.2.5   Displaying 2D Images Using image_view   78
          4.2.6   3D Data Visualization Using rqt_rviz (rviz)   80
          4.2.7   Recording and Playing Back Data Using rosbag and
                   rqt_bag   82
          4.2.8   rqt Plugins and rx Applications   84
   4.3   Summary of Basic ROS Commands   84
          4.3.1   Creating a ROS Workspace   85
          4.3.2   Package Related Operations   85
          4.3.3   Related Operations of Nodes   86
          4.3.4   Related Operations of the Topic   87
          4.3.5   Related Operations of the Service   88
          4.3.6   Related Operations of rosparam   89
          4.3.7   Bag-Related Operations   89
          4.3.8   Related Operations of rosmsg   90
          4.3.9   Related Operations of rossrv   91
          4.3.10   Other Commands of ROS   91

Part II   Implementation of Core Robot Functions

5   Installation and Initial Use of the Robots   95
   5.1   Introduction to the Turtlebot Robot   95
   5.2   Composition and Configuration of the Turtlebot Robot
          Hardware   96
   5.3   Installation and Testing of the Turtlebot Robot Software   98
          5.3.1   Installing from Source   98
          5.3.2   Deb Installation Method   99
          5.3.3   Configuration According to Kobuki Base   100
   5.4   Launching Turtlebot   101
   5.5   Manual Control of Turtlebot via Keyboard   102
   5.6   Controlling Turtlebot Through Scripting   103
   5.7   Monitoring the Battery Status of the Kobuki   104
   5.8   Extensions to the Turtlebot Robot   106
   Further Reading   108

6   Implementation of Robot Vision Functions   109
 6.1 Vision Sensors   109
   6.1.1 Kinect Vision Sensor   110
   6.1.2 Primesense Vision Sensors   111
 6.2 Driver Installation and Testing   112
 6.3 Running Two Kinects at the Same Time   113
 6.4 Running Kinect and Primesense at the Same Time   116
 6.5 RGB Image Processing With OpenCV in ROS   117
   6.5.1 Installing OpenCV in ROS   117
   6.5.2 Using OpenCV in ROS Code   118
   6.5.3 Understanding the ROS-OpenCV Conversion
      Architecture   119
   6.5.4 ROS Node Example   121
 6.6 Point Cloud Library and Its Use   125
   6.6.1 Introduction to Point Clouds and Point Cloud
      Libraries   125
   6.6.2 Data Types of PCL   125
   6.6.3 Publish and Subscrib to Point Cloud Messages   127
   6.6.4 Tutorial on How to Use PCL in ROS   129
   6.6.5 A Simple Application of PCL - Detecting the Opening
      and Closing State of a Door   136
 Further Reading   137

7   Advanced Implementation of Robot Vision Functions   139
 7.1 Implementation of the Robot Follow Function   139
   7.1.1 Theoretical Foundations   139
   7.1.2 Operational Testing of Follow Functions   140
 7.2 Implementation of the Robot Waving Recognition Function   144
   7.2.1 Implementation Framework and Difficulties Analysis
      of Robot Hand Waving Recognition   145
   7.2.2 Face Detection Based on AdaBoost and Cascade
      Algorithms   147
   7.2.3 Identifying Human Hands With the Template
      Matching Algorithm   149
   7.2.4 Skin Tone Segmentation Based on YCrCb Color
      Space   151
   7.2.5 Operational Testing of the Wave Recognition
      Function   152
 7.3 Implementation of Object Recognition and Localization
   Functions of the Robot   152
   7.3.1 Sliding Window Template Matching Method
      Based on Hue Histogram   152
   7.3.2 Object Localization Methods Based on Spatial
      Point Cloud Data   154
   7.3.3 Implementation and Testing of Object Recognition
      and Localization   156

7.4   Implementation of Face and Gender Recognition for Service
Robots. 157
    7.4.1   Traditional Face and Gender Recognition Methods
Based on OpenCV. 158
    7.4.2   Operational Testing of the OpenCV-Based Face
and Gender Recognition Function. 159
    7.4.3   Face Recognition Method Based on Dlib Library 160
    7.4.4   Operational Testing of the Face Recognition Function
Based on the Dlib Library. 161
7.5   Using TensorFlow to Recognize Handwritten Numbers. 167
    7.5.1   Introduction to TensorFlow. 167
    7.5.2   Installing TensorFlow. 167
    7.5.3   Basic Concepts of TensorFlow. 168
    7.5.4   Handwritten Digit Recognition Using TensorFlow. . . . 171
Further Reading. 175

8   Autonomous Robot Navigation Function. 177
8.1   Key Technologies for Autonomous Robot Navigation. 177
    8.1.1   Robot Localization and Map Building. 178
    8.1.2   Path Planning. 181
8.2   Kinematic Analysis of the Kobuki Base Model. 184
8.3   Navigation Package Set. 187
    8.3.1   Overview of the Navigation Package Set. 187
    8.3.2   Hardware Requirements. 187
8.4   Basics of Using the Navigation Project Package Set. 188
    8.4.1   Installation and Configuration of the Navigation
Project Package Set on the Robot. 188
    8.4.2   Robot tf Configuration. 196
    8.4.3   Basic Navigation Debugging Guide. 202
    8.4.4   Release of Odometer Measurements Via ROS. 205
    8.4.5   Publishing Sensor Data Streams Via ROS. 210
8.5   Configuring and Using the Navigation Project Package
Set on Turtlebot. 217
    8.5.1   Creating SLAM Maps Via Turtlebot. 217
    8.5.2   Autonomous Navigation Via Turtlebot's Known
Maps. 219
Further Reading. 220

9   Robot Voice Interaction Functions of Basic Theory. 223
9.1   Speech Recognition. 224
    9.1.1   Acoustic Models. 224
    9.1.2   Language Models. 229
9.2   Semantic Understanding. 233
    9.2.1   Seq2Seq. 233
9.3   Speech Synthesis. 235
Further Reading. 236

10 Implementation of Robot Voice Interaction Functionality: PocketSphinx 239
  10.1 Hardware 239
  10.2 Introduction of PocketSphinx Speech Recognition System 240
  10.3 Installing, Testing PocketSphinx on Indigo 242
    10.3.1 Installation of PocketSphinx 242
    10.3.2 Installing a Sound Library for Speech Synthesis 244
    10.3.3 Language Modeling with Online Tools 245
  10.4 Installing, Testing PocketSphinx on Kenetic 246
    10.4.1 Installing PocketSphinx on Kenetic versions 246
    10.4.2 Testing of PocketSphinx Speech Recognition 247

11 Implementation of Robot Arm Grasping Function 253
  11.1 Components of the Robot Arm Hardware 253
  11.2 Kinematic Analysis of the Robot Arm 254
  11.3 Setting the Servo ID of the Robot Arm 256
  11.4 Controlling Turtlebot-Arm with USB2Dynamixel 262
    11.4.1 Installing and Testing the dynamixel_motor
        Package 262
    11.4.2 Implementation of the Robotic arm Gripping
        Function 266

Part III Applications of Robots

12 Integrated Robots Application Case 1: Long Command Recognition and Multitasking Execution 273
  12.1 Case Objectives 273
  12.2 Voice Recognition Tasks 274
  12.3 Autonomous Navigation in the Home Environment 274
  12.4 Object Recognition and Grasping 277

13 Integrated Robots Application Case 2: Following and Assisting the User 279
  13.1 Case Objectives 279
  13.2 Voice Recognition Commands 281
  13.3 Following and Autonomous Navigation 281
  13.4 Detecting and Recognizing Faces 282

14 Integrated Robotics Application Case III: Customers Wave for the Robot Ordering 285
  14.1 Case Objectives 285
  14.2 Robot Real-Time Mapping 285
  14.3 Robot Recognizes Wavers and Moves to Waving Man 288
  14.4 Voice Recognition Menu 289
  14.5 Autonomous Navigation Back to the Bar 289

Correction to: Intelligent Robot C1

Part I
Basic Knowledge

The first part of the book will cover the basics of robots, including the following four chapters:

Chapter 1 Overview of Robots. Provide an introduction to the basic concepts of robots, including the definition, classification, development history, components, key technologies, and development trends of robots. Through this chapter, readers can have a clearer understanding and appreciation of current robots (especially service robots).

Chapter 2 Getting Started with ROS. It begins with an introduction to why ROS is used, the differences between robot operating systems and computer operating systems, and the main features of ROS; next, it describes the installation of ROS, including the version of ROS and the installation and uninstallation of both ROS Indigo and Melodic. The projects covered in this book are mainly run on ROS Indigo, and part of the programs are also tested on Melodic, the latest version of ROS. Finally web resources related to ROS learning are given.

Chapter 3 ROS Framework and Usage Fundamentals. the ROS Framework section covers the file system level, the computational graph level, and the community level; the ROS Usage Fundamentals section covers an introduction to catkin, workspaces and their creation, creating compiled project packages, creating compiled and running ROS nodes, the use of roslaunch, creating ROS messages and services, how to use C++ or Python to write Publishers and Subscribers for Testing Message, how to use C++ or Python to write Test Servers and Clients, etc.

Chapter 4 Debugging ROS. ROS provides a large number of commands and tools to help developers debug code in order to solve various software and hardware problems. This chapter introduces the commands and tools commonly used for ROS debugging and summarizes the basic ROS commands. Mastering these commands and tools is helpful for developers to solve the problems encountered in the process of robot debugging.

Chapter 1
Overview of Robots

This chapter provides an introduction to the basic concepts of robots, including the definition, classification, development history, components, key technologies, and development trends of robots. Through this chapter, readers can have a clear understanding and appreciation of robots (especially service robots).

1.1 Definition and Classification of Robots

1.1.1 Definition of Robots

The research of modern robots began in the mid-twentieth century, and the development of computers and automation and the development and utilization of atomic energy provided the technical basis for the research of modern robots. Since the mid-1980s, robots have entered people's daily lives from factories, and service robots that support people's daily lives have become an important direction in the development of robots. In recent years, with the rapid development of Internet technology, information technology and artificial intelligence technology, the experience of using service robots has been further enhanced. Through the deep integration of intelligent technologies such as automatic positioning and navigation, voice interaction and face recognition with robotic technology, people have developed a variety of new intelligent service robots, which has enabled the robot industry to usher in new opportunities for rapid development.

Depending on the intended application scenario, robots can be divided into two categories: industrial robots and service robots. There is no universally agreed international definition of robots, service robots, etc. Wikipedia uses the definition of robot from the Oxford English Dictionary (2016 edition):

A robot is a machine that can be programmed and can automatically perform a complex series of actions. A robot can be controlled by an external control device or

© The Author(s), under exclusive license to Springer Nature Singapore Pte Ltd. 2023
F. Duan et al., *Intelligent Robot*, https://doi.org/10.1007/978-981-19-8253-8_1

by a controller embedded within it. Robots can be constructed in human form, but most robots are machines that are used to accomplish tasks regardless of their form.

For service robots, Wikipedia uses the definition proposed by the International Federation of Robotics (IFR).

A service robot is a robot that can perform services beneficial to humans and equipment semi-automatically or fully autonomously, but excluding industrial manufacturing operations.

The International Organization for Standardization (ISO) also provides a definition of robots and service robots. ISO-8373-2012 defines robots and service robots as follows:

A robot is a drive with a degree of autonomy programmable for two or more axes, capable of moving through its environment to perform a predetermined task. In this context, "autonomy" denotes the ability to perform a predetermined task without human intervention, based on the current state and sensing. "A degree of autonomy" encompasses everything from partial autonomy (including human-machine interaction) to full autonomy (UAV operational intervention).

Service robots are robots that perform tasks that are useful to humans or devices, excluding industrial automation applications.

1.1.2 Classification of Service Robots

The International Federation of Robotics (IFR) classifies service robots according to their application areas and considers them to be divided into two categories: Personal/Domestic Service Robots and Professional Service Robots. ISO-8373-2012 gives a description of each of these two categories of robots:

A personal service robot is a type of service robot used for non-commercial tasks, usually for non-professionals.

Professional service robots are service robots used for commercial tasks and are usually operated by formally trained operators. In this context, an operator is a selected person who can start, monitor and stop the intended operation of the robot or robotic system.

Personal/Domestic service robots can be classified as home service robots, elderly assistance robots, edutainment robots, private automatic navigation vehicles, home security surveillance robots, etc.; professional service robots can be classified as venue robots, professional cleaning robots, medical robots, inspection and maintenance robots, construction robots, logistics robots, rescue and security robots, defense robots, underwater operation robots, powered human exoskeletons, commonly used drones, commonly used mobile platforms, etc.

1.2   The Evolution of Modern Robots

Modern robots research began in the mid-twentieth century and has been a rapidly developing field for more than half a century. This section will present the results of robotics development in different periods in chronological order.

1.2.1   Early Stages of Modern Robots Research

Early robots were mainly programmable robots, which were robots that could perform simple repetitive actions based on a program written by the operator.

In 1948–1949, William Grey Walter of the Burden Neurological Institute (England) designed the Machina Speculatrix, the first electronically autonomous robot that could perform complex movements. The robot was nicknamed "The Turtle" for its slow moving speed. It is a mobile robot with three wheels each powered by independent DC power, a light sensor, haptic sensor, propulsion motor, steering motor and two vacuum tubes to simulate a computer. The robot is phototropic and able to find charging posts to recharge. Walter named the first two robots Elmer and Elsie, and Fig. 1.1 shows Elsie without its housing.

In 1954, George Devol in the United States built the first digitally operated programmable robot and named it Unimate, thus laying the foundation for the

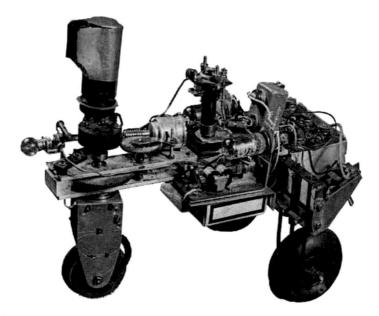

Fig. 1.1 Elsie without its housing

modern robotics industry. In 1960, Devol sold the first Unimate to General Motors for lifting hot metal parts from a die-casting machine and stacking those metal parts.

From 1962 to 1963, the development of sensor technology improved the maneuverability of robots. People began to experiment with various types of sensors on robots. For example, Ernst used tactile sensors in 1961; Tomovich and Boni installed pressure sensors on the world's first "dexterous hand" in 1962; McCarthy added a vision sensing system to the robot in 1963, and in 1964 helped the Massachusetts Institute of Technology (MIT) introduce the System/360, the world's first robotic system with vision sensors that could identify and locate blocks.

In 1968, Marvin Minsky built a computer-controlled, hydrodynamically driven 12-joint tentacle arm (Tentacle Arm); in 1969, mechanical engineering student Victor Scheinman invented the Stanford Robotic Arm, which is recognized as the first computer-controlled robotic arm (the arm's control commands are stored on magnetic drum).

From 1966 to 1972, the Center for Artificial Intelligence at the Stanford Research Institute (now RSI International) conducted research on the mobile robot system known as Shakey. In 1969, RSI published Shakey's "Robot Learning and Planning Experiments" video. The system has limited perception and environment modeling capabilities with multiple sensor inputs, including cameras, laser rangefinders and collision sensors. Shakey comes with vision sensors and can discover and grasp blocks on human command, and can perform tasks such as planning, pathfinding, and simple object rearrangement. However, the computer controlling it is the size of a room. Shakey is considered the world's first intelligent robot and kicked off the research on intelligent robots.

1.2.2   The 1970s

In the 1970s, robots gradually moved towards industrial applications, and they became progressively more perceptive and adaptive.

In 1970, the WABOT project was initiated by the Faculty of Science and Technology of Waseda University, Japan. Completed in 1973, WABOT-1 (shown in Fig. 1.2) is the world's first humanoid intelligent robot, including a hand and foot system, a vision system, and a sound system. It can communicate with humans in simple Japanese through an artificial mouth, measure the distance and direction of objects through artificial eyes and ears that act as remote receivers, move on foot with two feet, and grasp moving objects with hands that have tactile sensors.

In 1973, the German company KUKA developed the first 6-axis industrial robot, FAMULUS, driven by electric motors.

In the same year, the American company Cincinnati Milacron introduced the T3 robot, the first commercially available industrial robot controlled by a small computer.

Fig. 1.2 WABOT-1

In 1975, Victor Scheinman in the USA developed the Programmable Universal Manipulator Arm (PUMA.) In 1977, Scheinman sold his design to Unimation, who further developed the PUMA, which was later widely used in industrial production.

In 1979, the Stanford car (shown in Fig. 1.3) successfully navigated through a room full of chairs. It relied primarily on stereo vision to navigate and determine distances.

1.2.3   The 1980s

In the 1980s, industrial robots were widely used, and service robots that support people's daily lives began to come into the public eye.

In 1981, Takeo Kanade of Japan developed the first "Direct-Drive (DD) arm" in which the drive for the arm was included in the robot, eliminating long distance

Fig. 1.3 Stanford car

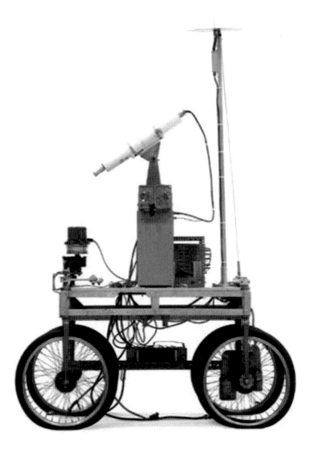

drives. This direct-drive arm is a prototype of the DD arm used in industry today. The motor mounted in the joint eliminates the need for chains or tendons used in earlier robots, and because friction and backlash are reduced, the DD arm operates quickly and accurately.

In 1982, the American company Heathkit released the educational robot HERO (Heathkit Educational RObot). HERO-1 (shown in Fig. 1.4) is a self-contained mobile robot controlled by an on-board computer with a Motorola 6808 CPU and 4KB RAM. The robot has light, sound, and motion detectors and a sonar range sensor that enables it to move through hallways, play games, sing, and even act as an alarm clock by using sonar navigation to find humans and keep following them by sound.

In 1982, Joseph Bosworth, the founder of RB Robotics in the United States, released the RB5X (shown in Fig. 1.5). The RB5X robot is the first mass-produced programmable robot for domestic, experimental, and educational use. The RB5X contains components such as infrared sensing, ultrasonic sonar, remote audio/video transmission, eight sensors/buffers, a voice synthesizer, and a 5-axis armature. The

Fig. 1.4 HERO-1

Fig. 1.5 Joseph Bosworth and RB5X

Fig. 1.6 WABOT-2

RB5X can play interactive games with up to eight players. Its program can be written and downloaded from a computer.

In 1984, WABOT-2 (shown in Fig. 1.6) was released. WABB-2 is not a general-purpose robot like WABB-1, but a robot that pursues ingenious artistic activities in the daily work of human beings. WABOT-2 can converse naturally with humans in Japanese, recognize music with its eyes, play the electronic piano with both hands and feet, and even play tunes of intermediate difficulty. It is also able to recognize human singing and perform automatic score picking so that it can accompany human singing. This means that the robot has the ability to adapt to humans, a big step towards personal robots.

In 1988, Gay Engelberger of the United States introduced HelpMate, the first assistive service robot designed for hospitals and sanatoriums. It consists of two products, HelpMID and HelpMead. HelpMID is an autonomous robot that uses vision, ultrasound and infrared to sense its environment and can navigate along corridors and avoid obstacles. HelpMead can navigate through hospitals, track maps stored in its memory, carry medical supplies, dinner trays, records and lab samples, and deliver them to nursing wards or other departments.

Fig. 1.7 Aquarobot, an
underwater walking robot

In 1989, the Robotics Laboratory of the Ministry of Transport of Japan developed Aquarobot, an underwater walking robot (shown in Fig. 1.7). The robot is a six-legged articulated "insect-type" walking fully automatic intelligent robot that can operate at depths up to 50 m. Aquarobot has an ultrasonic conversion system, which is a long-baseline navigation device. At the operator's end there is an underwater TV camera with an ultrasonic distance measuring device. The robot has two main functions: one is to measure the levelness of the rocky foundation of the breakwater by the movement of its legs while walking, and the other is to observe underwater structures using the camera.

1.2.4   The 1990s

In the 1990s, robotics developed more rapidly, with service robots in particular seeing rapid growth.

Founded in 1990 by Rodney Brooks, Colin Angle and Helen Greiner, the American company iRobot produces robots for domestic and military use.

In 1996, David Barrett of MIT invented the bionic robot RoboTuna to study how fish swim in water.

In the same year, Honda introduced a two-handed, bipedal humanoid robot, P2 (shown in Fig. 1.8). The P2 robot, which is 182 cm tall, 60 cm wide, and 210 kg in weight, is the world's first robot that can control itself and walk on two legs.

In 1997, Honda also developed the humanoid robot P3 (shown in Fig. 1.9). P3 was the first fully autonomous humanoid robot developed by Honda.

In 1998, the world's first bionic arm (shown in Fig. 1.10) was successfully matched to the Campbell Aird at the Princess Margaret Rose Hospital in England.

Fig. 1.8 P2

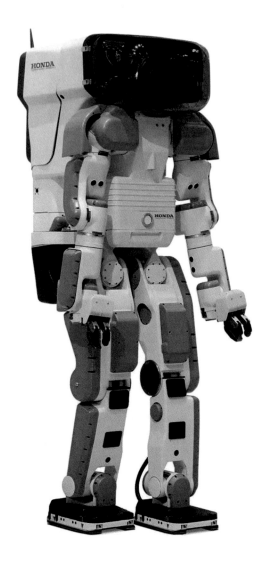

This arm was the first robotic arm consisting of a powered shoulder, elbow, wrist, and fingers controlled by electronic microsensors.

In 1998, the University of Tokyo and the National Institute of Advanced Industrial Science and Technology (AIST) developed the HRP-2 robot (shown in Fig. 1.11) and held a "life show" at the University of Tokyo, in which the robot was able to take away a cup of tea that a person had finished drinking, pour bottled tea into a cup of water, and wash water glasses.

In 1999, Sony introduced a robot dog called AIBO (shown in Fig. 1.12). The AIBO robot dog was equipped with a distance sensor on its chest, which increased joint flexibility and enabled lateral movement; with the addition of 28 LEDs in its head, the AIBO robot dog could express its emotional changes. The AIBOERS-7

Fig. 1.9 P3

Fig. 1.10 Bionic arm

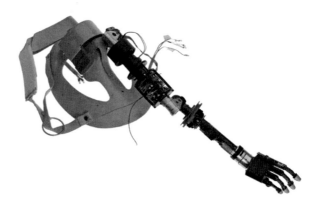

uses a 576MHz 64-bit processor with 64MB SDRAM memory, a 350K pixel CMOS camera as its "eyes", and a Infrared distance sensor, memory acceleration sensor, vibration sensor, electrostatic sensor, and many other sensing "organs". In addition, it has added wireless communication capability, supporting wireless LAN (IEEE 802.11b/Wi-Fi standard).

In 1999, Personal Robots released the Cye robot (shown in Fig. 1.13), the first practical domestic robot. It could find and dock with a charger, could transport things from one room to another, and had the function of a vacuum cleaner.

Fig. 1.11 The University of Tokyo's domestic service robot

Fig. 1.12 Sony's AIBO
robot dog

Fig. 1.13 Cye robot base cell and vacuum cleaner

Fig. 1.14 Roomba, a cleaning robot from iRobot USA

1.2.5   Twenty-First Century

In the twenty-first century, Internet technology, information technology and artificial intelligence technology have developed rapidly, the intelligence level of robots has improved rapidly, and there has been a series of breakthroughs in intelligent robotics, and service robots are gradually going to thousands of households.

In 2000, Honda presented its most advanced achievement in a humanoid project. The robot was named Asimo. Asimo can run, walk, communicate with people, recognize faces, environment, sounds and postures, and interact with its environment. Honda has also introduced ASIMO, a bipedal robot. ASIMO is 130 cm tall and weighs 54 kg, and it can walk up to 3 km/h, almost the same as a human. It can also run at 6 km/h and gyrate at 5 km/h with a radius of 2.5 m. ASIMO has preliminary artificial intelligence and can make movements according to predetermined settings, as well as respond to human voices and gestures, and has basic memory and recognition capabilities.

In the same year, Sony released a small bipedal entertainment robot, the SDR-3X (Sony Dream Robot), which has good motor performance.

In 2001, the Canadarm2 robotic arm was launched into orbit and attached to the International Space Station. The arm, developed in Canada and carried aboard the United States space shuttle Endeavour, is longer, stronger and more agile than the smaller arms that flew with the space shuttle on previous missions.

In the same year, the Global Hawk drone made the first autonomous non-stop flight over the Pacific Ocean from Edwards Air Force Base in southern California to Edinburgh Air Force Base in southern Australia.

In 2002, iRobot introduced the first generation of its indoor carpet and floor cleaning robot, Roomba (shown in Fig. 1.14), which automatically plans routes, avoids obstacles, and returns to the charging cradle to recharge when the battery is low. Roomba is currently the world's largest selling and most commercially available household robot.

In the same year, InTouch Health, a U.S. company focused on the combination of robotics and the Internet, was founded. It introduced a telemedicine robot (shown in Fig. 1.15) with a real-time two-way video and voice delivery system and a walking

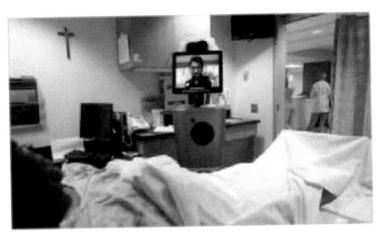

Fig. 1.15 Telemedicine Robot from InTouch Health, USA

system to facilitate remote communication between doctors and patients. Robotic security guards with various detection devices are also popular with the U.-S. military's logistics department and major companies.

In 2003, NASA launched twin robotic Mars Exploration Rovers to Mars to find answers about the history of water on Mars.

In 2004, Cornell University researched a robot capable of self-replication (shown in Fig. 1.16), a set of cubes capable of attachment and disassembly, and the first robot capable of making its own copies.

In 2004, Paro (shown in Fig. 1.17), a seal-type pet robot with an advanced intelligent system developed by the Japan Advanced Institute of Industrial Technology (AIST), was launched. It is extremely sensitive to people's actions. According to people's different performance, it will make different reactions such as joy and anger. The main feature of Paro is that it can respond interactively to human touch for the purpose of comforting the human and thus serving a therapeutic purpose. In its company, sick children and lonely elderly people will get a lot of joy. For this reason, it has been called "the world's most therapeutic robot" by Guinness World Records.

In 2004, the irobi home robot (shown in Fig. 1.18) was introduced in South Korea. It has a home security function that helps confirm that the door is locked and the gas is off when the house is unoccupied. It can also photograph the intruder if someone breaks into the house and send a photo to the owner via e-mail. In addition, irobi can read Korean and English books and can sing nursery rhymes as well as tell fairy tales, thus becoming a new partner for children's education.

In 2006, Cornell University demonstrated its development of the "Starfish" robot (shown in Fig. 1.19), a quadrupedal robot that can model itself and recover the ability to walk after being damaged.

In 2007, TIPS launched the entertainment robot i-sobot, the world's smallest bipedal walking robot at the time, which could walk like a human, perform kicking

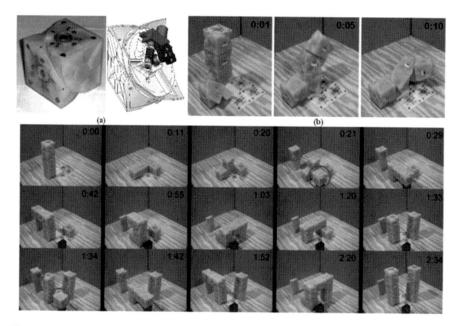

Fig. 1.16 Cornell University's reproducible robot

Fig. 1.17 A seal-type pet
robot from Japan, Paro

Fig. 1.18 Irobi robot

and boxing moves in special movement mode, and play fun games and perform difficult special moves.

In the same year, researchers at Waseda University in Japan demonstrated Twendy-One, a new humanoid robot dedicated to the elderly and disabled (shown

Fig. 1.19 "Starfish" robot

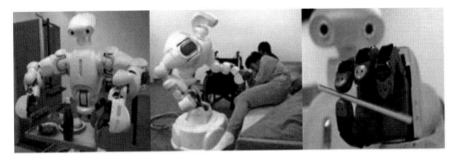

Fig. 1.20 Twendy-One service robot from Waseda University, Japan

in Fig. 1.20). The robot was able to carry a man out of bed, talk to him, and prepare breakfast for him. It is also able to remove bread from a toaster with a clip without breaking it at all, and then place it on a tray with a bottle of ketchup from the refrigerator and hand it to the man. Even more amazing is that Twendy-One can easily grab a straw with its fingers.

In 2010, Willow Garage USA introduced the PR2 personal service robot (shown in Fig. 1.21). The PR2 is an open ROS-based platform capable of serving humans in a home or work environment, and users can change the system to meet their needs. The PR2 not only has the mobility to navigate autonomously through the environment, but also has the flexibility of grasping and manipulating objects to perform tasks such as clearing tables, folding towels, and getting drinks from the refrigerator.

In the same year, the Korea Institute of Science and Technology (KIST) developed Mahru-Z (shown in Fig. 1.22), a 1.3 m tall, 55 kg humanoid robot with six

Fig. 1.21 PR2 robot
developed by Willow
Garage

Fig. 1.22 Mahru-Z robot from KIST, Korea

fingers, a rotating head, and three-dimensional vision. Mahru-Z can perform a variety of household tasks, including cleaning the house, putting clothes in the washing machine, and heating food in the microwave, and it can distinguish between different people. KIST invested nearly 4 billion won and took 2 years to complete the study. Before it is officially put into commercial use, KIST plans to further

a) Home Service Robot HSR   b) HSR moves the bracket and arm simultaneously

Fig. 1.23 HSR, a home service robot developed by Toyota Motor Corporation, Japan. (**a**) Home Service Robot HSR (**b**) HSR moves the bracket and arm simultaneously

enhance the robot's skills so that it can perform tasks such as cooking, washing dishes and serving its owner.

In 2011, NASA delivered Robonaut 2, a new generation of astronaut assistants, to the International Space Station aboard the Space Shuttle Discovery. Robonaut 2 is the first humanoid robot to operate in space, and although its primary job is to teach engineers how to operate robots in space, it is hoped that through upgrades and advancements it will 1 day be able to leave the station to help astronauts with missions such as spacewalks, maintenance and scientific experiments.

In 2012, Toyota Motor Corporation of Japan released the Home Service Robot HSR (Human Support Robot) (shown in Fig. 1.23a), but it was not sold. The software architecture of HSR is built on top of ROS. The robot is widely used among robotics researchers as an open platform. The HSR robot is fitted with multiple sensors and has 8 degrees of freedom overall, including 3 degrees of freedom for moving the base, 4 degrees of freedom for the arm, and 1 degree of freedom for torso lifting. Thus, flexible motion can be accomplished by moving the bracket and arm simultaneously (as shown in Fig. 1.23b). HSR is a mobile robotic robot with both physical labor and communication capabilities. HSR is capable of picking up and carrying objects, and its development goal is to be able to perform tasks such as operating furniture (e.g., opening/closing drawers, using a microwave oven, etc.), picking up and carrying everyday objects, and organizing a room.

In recent years, with the rise of technologies such as big data, cloud computing, machine learning, deep learning, and artificial intelligence, and their wide application in computer vision, natural language processing, and autonomous navigation, the development of robotics, especially service robots, has been greatly facilitated.

In 2014, the South Korean company Future Robot introduced a hotel service robot called FURO (shown in Fig. 1.24). This robot has a beauty avatar on its head display that mimics various expressions. FURO is equipped with multiple sensors

Fig. 1.24 A hotel service robot, FURO

that can sense people and obstacles. FURO holds a large display that makes it easy to make hotel reservations and has a swipe card spending function. In addition to being used for hotel reservations, FURO can also be used in shops and airports to navigate people.

In the same year, Fraunhofer IPA of Germany released the Care-O Bot 4 series of mobile service robots, shown in Fig. 1.25b, designed to bring service robots into the home or commercial space. It is more sophisticated than Care-O Bot 1, built as a prototype in 1998, Care-O Bot 2, developed in 2002, and Care-O Bot 3, developed in 2010 (shown in Fig. 1.25a). Care-O Bot 4 not only serves as a pickup and handling assistant, but also has communication and entertainment functions. It has a sensor-driven navigation system to reach its destination via the optimal route. The enhanced flexibility of the Care-O Bot 4 is due to the design of spherical joints around its neck and hip pivot points, which extend the robot's working space and allow the head and torso to rotate 360°. Care-O Bot 4 reduces development costs and is more adaptable to its environment. It is capable of detecting and overcoming obstacles through sensors, and the technologies that enable rapid response include 3D sensors, laser

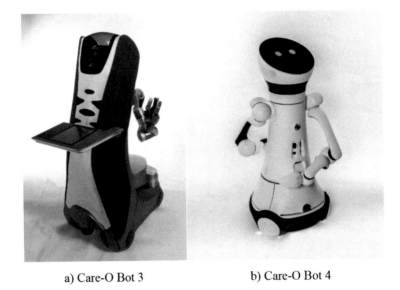

a) Care-O Bot 3   b) Care-O Bot 4

Fig. 1.25 Care-O bot series robots developed by Fraunhofer IPA, Germany. (**a**) Care-O Bot 3 (**b**) Care-O Bot 4

scanners and stereo vision cameras. Through sensors, the robot is also able to identify objects on its own and perform 6D calculations on their location.

In addition, in 2014, Japan's SoftBank Group introduced the Pepper robot (shown in Fig. 1.26), the world's first social humanoid robot capable of recognizing faces and basic human emotions. The Pepper robot stands 120 cm tall and is capable of communicating with people through dialogue and a touch screen. Pepper is an open, fully programmable platform, whose head has speech recognition and dialogue functions in 205 languages, including English, French, German, Spanish, Arabic, Italian, Dutch, etc. It has a perception module that recognizes and interacts with the person it is talking to. It has touch sensors, LEDs and microphones for multi-modal interaction, as well as infrared sensors, buffers, inertial devices, 2D and 3D cameras and sonar for omni-directional and autonomous navigation.

In May 2014, Microsoft Asia Internet Engineering Institute launched the artificial intelligence companion virtual robot "Microsoft Ice", which is based on Microsoft's emotional computing framework, integrated use of algorithms, cloud computing and big data technology, through the intergenerational upgrade method, and gradually formed a complete artificial intelligent system towards EQ direction.

In October 2017, a robot named Sophia was granted Saudi citizenship at the Future Investment Summit in Riyadh, becoming the first robot ever to have a nationality, but causing much controversy.

In summary, in the mid-1980s, robots began to move from factory environments into people's everyday environments, including hospitals, offices, homes, or other cluttered and uncontrollable environments. This required robots that could not only

Fig. 1.26 Japan's SoftBank
Group introduced the
Pepper robot

do the job autonomously, but also collaborate with or be guided by a human to complete the task. In recent years, in particular, one can see the emergence of a variety of autonomous mobile robots that will clean floors, mow lawns, or act as guides, nannies, and guards, among other things.

1.3   Composition of the Robots

Robots generally consist of actuators, drives, sensing devices, control systems, and complex machinery. This section will describe each of these components of an intelligent service robot.

1.3.1 Actuators for the Robot

The actuator of a robot is the body of the robot, such as the robot arm and the walking mechanism that make up the robot. The arm of the robot (if any) is usually a spatial open chain linkage mechanism, and the motion subsets (rotating or moving subsets) on it are generally called joints, and the number of joints is generally referred to as the number of degrees of freedom of the robot. Depending on the form of joint configuration and the form of motion coordinates, robot actuators can be classified into right-angle coordinate, polar coordinate, cylindrical coordinate, and joint coordinate types. For certain application scenarios, based on anthropomorphic considerations, the relevant parts of the robot body are usually referred to as the base, waist, arm, wrist, hand (gripper or end-effector) and walking parts (for mobile robots), etc.

1.3.2 Drive Units

The drive unit is a device that drives the motion of the actuator, which can drive the robot to perform the corresponding action with the help of power elements according to the command signal from the control system. In general, the drive receives an electrical signal input and produces a linear or angular displacement output. Robots usually use electric drives, such as stepper motors, servo motors, etc. In addition, for the specific needs of a particular scenario, hydraulic and pneumatic drives can also be used.

1.3.3 Sensing Devices

The robot obtains information from the outside world through various sensors. Sensors are generally used to monitor the robot's internal motion, operating conditions, and information about the external working environment in real time, which is then fed back to the control system. The control system processes the feedback information to drive the actuators and ensure that the robot's movements meet the predetermined requirements. The sensors with detection role can be divided into two kinds: one is the internal information sensor, which can detect the internal conditions of the robot, such as the position, speed and acceleration of the joints, and feedback the measured information to the controller to form a closed-loop control; the other is the external information sensor, which is used to obtain the current operating object of the robot and the external environment and other information, so that the robot's action can adapt to the external changes and achieve a higher level of automation, or even make the robot have a certain human-like "feeling" and achieve intelligence. For example, external sensors such as vision and sound can obtain information about the work object and work environment, and feed this information to the control system to adjust the actuator and improve the work accuracy of the robot.

1.3.4   Control Systems

A control system generally refers to the computer that executes the control program. There are usually two types of control systems: a centralized control system, where a single microcomputer does all the control work of the robot; and a decentralized (level) control system, where multiple microcomputers share the control tasks of the robot, for example, by using two levels of microcomputers, upper and lower, to control together; the upper master is usually responsible for system management, communication, kinematic and dynamical calculations, and sending command information to the lower master. The lower level slaves (even up to one management slave per joint) are usually responsible for interpolation operations and servo control processing, implementing specific motion functions, and providing feedback to the host.

1.3.5   Intelligent Systems

An intelligent system is a computer system that produces human-like intelligence or behavior. The meaning of intelligence is very broad, the concept itself is constantly evolving, and the nature of intelligence needs to be further explored, so it is difficult to give a complete and precise definition of the term "intelligence", but the following expression is generally used: intelligence is the embodiment of the higher activities of the human brain, which should at least have the ability to automatically acquire and apply knowledge, think and reason, solve problems, and learn automatically. The "intelligence" of an intelligent service robot refers to the ability to perform functions similar to human intelligence. The main characteristic of an intelligent system is that it processes not only data but also knowledge. The ability to represent, acquire, access and process knowledge is one of the main differences between an intelligent service robot system and a traditional mechanical system. Therefore, an intelligent system is also a knowledge-based processing system, which requires the following technologies as a basis: a knowledge representation language; knowledge organization tools; methods and environments for building, maintaining and querying knowledge bases; and support for reuse of existing knowledge. Intelligent systems typically use artificial intelligence problem solving models to obtain results. Compared to the solution models used by traditional systems, artificial intelligence solution models have three distinctive features, namely, their problem solving algorithms are often non-deterministic or heuristic; their problem solving relies heavily on knowledge; and the problems to be solved by intelligent systems tend to have exponential computational complexity. The problem solving methods used by intelligent systems are generally classified into three categories: search, inference, and planning. Another important difference between intelligent service robot systems and traditional systems is that intelligent systems have field-awareness (environment adaptation) capabilities. By site perception we mean that the robot can interact with an abstract prototype of the real world it is in and adapt to the live environment it is in. This interaction usually involves perceiving, learning, reasoning, judging and making appropriate actions, which is often referred to as self-organization and self-adaptability.

1.3.6 Intelligent Human-Machine Interface Systems

Intelligent service robots cannot be fully autonomous at present, they still need to interact with humans, and even fully autonomous robots need to provide real-time feedback to humans on task execution. An intelligent human-machine interface system refers to a more friendly and natural human-machine interaction system with good adaptive capabilities provided by the robot to the user. An intelligent HMI system should include the following features: the ability to conduct direct human-robot dialogue in natural language, the ability to interact with humans through a variety of media such as sound, text, graphics and images, and even the ability to interact with humans through physiological signals such as brain waves, adapting to different user types, different needs of users, and the support of different computer systems.

1.4 Key Technologies for Robots

Robot is a highly flexible automated mechanical system capable of performing a variety of tasks in place of people. Robotics is a highly integrated new technology that brings together a variety of disciplines such as mechanics, electronics, mechanology, biology, artificial intelligence, cybernetics, and system engineering. Service robotics is similar in nature to other types of robots.

1.4.1 Autonomous Mobile Technologies

One of the key technologies for service robots is autonomous mobile technology, the most important of which is robot navigation technology. Service robots achieve the goal of autonomous movement (i.e., navigation) by automatically avoiding obstacles in the environment by detecting the environment and sensing their own state through sensors. Commonly used indoor robot navigation technologies include magnetic navigation, RFID navigation, ultrasonic and radar navigation, voice navigation and visual navigation.

1.4.2   Perceptual Technologies

The sensing system of a service robot is a sensing network consisting of various sensors together. These include pressure sensors for sensing touch, smoke and hazardous gas sensors for sensing the indoor environment, photoelectric sensors for sensing the intensity of light in the room, speed sensors for measuring movement speed and distance, proximity sensors for precise movement positioning over short distances, and voice sensors for enabling human-robot dialogue and completing voice commands. In addition, there are vision sensors that sense the indoor spatial environment, which allow the robot to better perform functions such as object recognition, positioning and grasping. Currently, voice sensing technology and spatial vision sensing technology are widely researched and applied in the field of service robots, enabling service robots to complete humanoid environment perception through voice and vision.

1.4.3   Intelligent Decision-Making and Control Technologies

Intelligent decision and control refers to automatic decision control technology that autonomously drives intelligent machines without human intervention, makes intelligent decisions and achieves control goals. Intelligent decision control of service robots includes automatic and intelligent decision making and control in the process of autonomous movement, precise positioning, recognition of grasping objects, human-machine interaction, network control, etc. Intelligent decision control contains: fuzzy control, neural network control, artificial intelligence control, humanoid control, chaos control, etc. Intelligent decision and control techniques are mainly used to solve complex control situations where the control object cannot be accurately modeled and has characteristics such as nonlinearity.

1.4.4   Communication Technologies

Service robots mainly apply communication technologies to realize the transmission of data such as sensing, sound and images over the network, and can also receive commands and controls from remote networks. Current remote interaction methods for service robots include GSM data transmission communication based on cellular networks, TCP/IP-based wired and wireless network communication, and wireless sensing network (WSN)-based interaction communication.

1.5 Trends in Robots

The previous sections have analyzed the history and important technologies of service robots, and this section will provide an outlook on the development trends of service robotics.

1.5.1 Hierarchical and Humanized Human-Computer Interaction

With the development of a new generation of service robots, the interaction between the robot and the user shows a trend of hierarchy. For example, the user can give relatively high-level instructions to the robot, and the robot can even guess the user's needs through automatic recognition; the user can also give relatively low-level instructions directly to the robot, such as having the robot follow the user's movements completely. The user can communicate directly with the robot at close range, and can also interact remotely through communication channels such as the Internet. The interaction between users and service robots will also become increasingly humanized; users will be able to interact with robots in natural and intuitive ways such as through voice and gestures, and robots will be able to provide feedback to users in more diverse ways.

1.5.2 Interaction Intelligence with the Environment

As sensor technology and processes advance, the next generation of service robots will also be equipped with more diverse and advanced sensing devices, allowing for more accurate identification of objects in the environment and more precise judgments about the state of the environment. Service robots can also operate objects in the environment in a more refined manner, thus providing users with a richer range of service functions. The entry of robots into the home has not only accelerated the development of robots themselves, but has also led to the continuous improvement of the home environment. By intelligently modifying the robot's working environment, the functions of the robots working in it can also be enhanced. In the future, service robots and the home environment will contribute to each other, forming a closely integrated organic intelligence.

1.5.3 Networking of Resource Utilization

The combination of the Internet and robots is an important development direction for service robots. The Internet is like a huge resource bank with huge computing and information resources, and the combination of robots and the Internet is a means to effectively use these resources. As an intelligent terminal and operation carrier, the service robot itself has the functions of movement, perception, decision making and operation. With the cloud computing, big data, Internet of Things and other technologies of the Internet platform, service robots can obtain an information collection and processing platform with great potential. Combined with the Internet, it can largely extend the perception, decision-making and operation capabilities of service robots.

1.5.4 Standardization, Modularization and Systematization of Design and Production

With the increasing development of service robots, the establishment of a widely recognized service robot standard, the design and establishment of a modular architecture for service robots has become an urgent problem in the development of service robots. Promoting the standardization, modularization and systematization of service robot design and production can reduce repetitive labor, speed up the transformation of advanced technologies into products, improve the quality of service robot products, reduce costs and promote the industrialization of service robots.

Exercises
1. Please list the robots you have come into contact with?
2. What are the key technologies included in modern robotics?
3. What trends do you think will happen with robots?

Further Reading

1. Peter Kopacek. Development Trends in Robotics [J]. E&I Elektrotechnik Und Informationstechnik, 2013, 130(2): 42–47.
2. The Oxford English Dictionary [Z]. Oxford University Press, 2016.
3. Michael Gasperi's Extreme NXT. Machina Speculatrix [EB/OL]. http://www.extremenxt.com/walter.htm.
4. Patrick Waurzyniak. MASTERS OF MANUFACTURING: Joseph F. Engelberger[J]. Society of Manufacturing Engineers, 2006, 137(1): 65–75.
5. Artificial Intelligence Center. shakey [EB/OL]. http://www.ai.sri.com/shakey/.
6. Computer History Museum. Timeline of Computer History [EB/OL]. http://www.computerhistory.org/timeline/ai-robotics/.
7. Wikipedia. HERO [EB/OL]. https://en.wikipedia.org/wiki/HERO_(robot).
8. Robot Workshop. RB5X-RB Robotics [EB/OL]. http://www.robotworkshop.com/robotweb/?page_id=122.
9. cyberneticzoo.com. 1982 - RB5X the Intelligent Robot - Joseph Bosworth (American) [EB/OL]. http://cyberneticzoo.com/robots/1982-rb5x-the-intelligent-robot-joseph-bosworth-american/.

10. humanoid. wabot: wAseda roBOT [EB/OL]. http://www.humanoid.waseda.ac.jp/booklet/
    kato_2-j.html.
11. JohnEvans, BalaKrishnamurthy, WillPong, et al. HelpMate: A robotic materials transport
    system [J]. Robotics and Autonomous Systems, 1989, 3:251-256.
12. cyberneticzoo.com. 1985 - 无缝quarobot?Aquatic walking robot - (Japanese) [EB/OL]. http://
    cyberneticzoo.com/underwater-robotics/1985-aquarobot-aquatic-walking-robot-japanse/.
13. Roomba. roomaba i series [EB/OL]. http://www.irobot.cn/brand/about-irobot.
14. K Hirai, M Hirose, Y Haikawa, et al. The development of Honda humanoid robot
    [J]. IEEE, 2002.
15. National Museums Scotland. The first bionic arm [EB/OL]. https://www.nms.ac.uk/explore-
    our-collections/stories/science-and-technology/made-in-scotland-changing-the-world/scottish-
    science-innovations/emas-bionic-arm/?item_id=.
16. Love Academia. Robotics HRP-2 Grand Close-up [EB/OL]. https://www.ixueshu.com/
    document/86700b6ff3e385e318947a18e7f9386.html#pdfpreview.
17. CYE Robot. A Robot is knocking on your Door [EB/OL]. http://www.gadgetcentral.com/cye_
    robot.htm.
18. Wikipedia. History of robots [EB/OL]. https://en.wikipedia.org/wiki/history_of_robots#cite_
    note-63.
19. Baidu Encyclopedia. ASIMO [EB/OL]. https://baike.baidu.com/item/ASIMO/1312513?fr=
    aladdin.
20. Baidu.org. International Space Station [EB/OL]. https://baike.baidu.com/item/%E5%9B%BD
    %E9%99%85%E7%A9%BA%E9%97%B4%E7%AB%99/40952?fr=Aladdin.
21. Wikipedia. Mobile Servicing System [EB/OL]. https://en.wikipedia.org/wiki/Mobile_Servic
    ing_System#Canadarm2.
22. Douban. Interview with Youlun Wang, father of surgical and telemedicine robots [EB/OL].
    https://www.douban.com/note/525739320/
23. CREATIVE MACHINES LAB-COLUMBIA UNIVERSITY. MACHINE SELF REPLICA-
    TION [EB/OL]. https://www.creativemachineslab.com/self-replication.html.
24. TWENDY CHEM, Sugano Research Laboratory, Waseda University. Twendyone [EB/OL].
    http://www.twendyone.com/index_e.html.
25. Willow Garage. pr2 overview [EB/OL]. http://www.willowgarage.com/pages/pr2/overview.
26. IEEE ROBOTS. pr2 [EB/OL]. https://robots.ieee.org/robots/pr2/.
27. NASA. Robonaut 2 Getting His Space Legs [EB/OL]. https://www.nasa.gov/mission_pages/
    station/main/robonaut.html.
28. Hibikino-Musashi@Home. Human Support Robot (HSR) [EB/OL]. http://www.brain.kyutech.
    ac.jp/~hma/wordpress/robots/hsr/.
29. TOYOTA. PARTNER ROBOT [EB/OL]. https://www.toyota-global.com/innovation/partner_
    robot/robot/.
30. Projection Times.com. FURO robot to be presented at Shanghai HD Display and Digital
    Signage Technology Exhibition [EB/OL]. http://www.pjtime.com/2014/5/232519296129.
    shtml.
31. Fraunhofer. Care-O-bot 4 [EB/OL]. https://www.care-o-bot.de/en/care-o-bot-4.html.
32. ROBOTICS TODAY. Care-O-bot Series: Care-O-bot 4 [EB/OL]. https://www.roboticstoday.
    com/robots/care-o-bot-4-description.
33. ROBOTICS TODAY. Care-O-bot Series: Care-O-bot 3 [EB/OL]. https://www.roboticstoday.
    com/robots/care-o-bot-3-description.
34. SoftBank Robotics. pepper [EB/OL]. https://www.softbankrobotics.com/emea/en/pepper.
35. Ren FJ, Sun X. Status and development of intelligent robots [J]. Science and Technology
    Herald, 2015, 33(21):32–38.
36. JI Pengcheng, SHEN Huiping. The current situation of service robot and its development trend
    [J]. Journal of Changzhou University, 2010, 22(2):73–78.
37. Luo Jian. Current status and key technologies for the development of elderly service robots
    [J]. Electronic Testing, 2016(6):133–134.
38. Liang RJ, Zhang T, Wang XQ. A review of home service robots [J]. Smart Health, 2016, 2(2):
    1–9.

Chapter 2
Getting Started with ROS

In the field of robotics, the role of Robot Operating System (ROS) is similar to that of Android or iOS in the field of smartphones. The ROS provides a unified platform for robotics development, where more users can easily research and verify robotics algorithms and develop robotics applications, thus greatly facilitating the development of robotics.

Robot operating systems are of strategic importance to the development of the intelligent robotics industry, and countries around the world are currently competing to carry out research on robot operating systems. For example, Italy has developed the open-source robot operating system YARP, and Japan's National Institute of Advanced Industrial Science and Technology has developed the open-source robotics middleware OpenRTM-aist. The United States has invested more in this area, developing the well-known robot development platform ROBOTIES, Player Stage, and the widely used ROS (Robot Operating System).

The ROS open-source robot operating system was released by Willow Garage in 2010 and has quickly grown to become one of the mainstream robot operating systems because of its ease of use.

This chapter begins with a brief introduction to the robot operating system, including why you should use ROS for robot development, the concept of ROS, the differences between ROS and computer operating systems, and the main features of ROS. Then, the installation and configuration of ROS is explained in detail, including an introduction to ROS versions and how to install, configure, and uninstall both ROS Indigo and Melodic versions. The sample programs in this book are mainly run on the ROS Indigo version, but some of them are also tested on the Melodic version of ROS. Finally some web resources for learning ROS are listed.

© The Author(s), under exclusive license to Springer Nature Singapore Pte Ltd. 2023
F. Duan et al., *Intelligent Robot*, https://doi.org/10.1007/978-981-19-8253-8_2

2.1 Introduction to ROS

2.1.1 Why Use ROS

Robotics software development has many common issues, such as software usability, programming development efficiency, cross-platform development capabilities, multi-programming language support capabilities, distributed deployment, code reuse, etc. Anyone who has worked in robotics software development knows that designing and developing truly robust general-purpose robotics software is not an easy task. This is because problems that are trivial to humans tend to vary wildly for robots as their environments and tasks change. Dealing with these changes is very difficult and no one person, lab or institution can handle it alone.

In 2010, Willow Garage released ROS, an open-source robotics operating system, with the goal of solving the above problems. The ROS system was originally built on the work of different individuals, teams and labs, designed for collaboration with each other.

2.1.2 What Is ROS

As mentioned earlier, ROS is an open-source operating system for robots that provides the services expected of an operating system, including hardware abstraction, control of underlying devices, implementation of common functions, inter-process messaging, and package management. It also provides the tools and library functions needed to acquire, compile, write, and run code across computers. In some ways, ROS is the equivalent of a "robot framework", similar to Player, YARP, Orocos, CARMEN, Orca, MOOS, and Microsoft Robotics Studio.

The ROS runtime "blueprint" is a loosely coupled peer-to-peer process network based on the ROS communication infrastructure. ROS implements several different communicators, including service based on synchronous RPC style communication, topic based on asynchronous streaming media data and Parameter Server for data storage.

ROS is not a real-time framework, but ROS can be embedded in real-time programs. Willow Garage's PR2 robot uses a system called pr2_etherCAT to send or receive ROS messages in real-time. ROS can also be seamlessly integrated with the Orocos real-time toolkit.

2.1.3   Differences Between ROS and Computer Operating Systems

According to the Wikipedia definition, OS is system software that manages computer hardware and software resources and provides common services for computer programs. ROS is also an operating system (OS), which has similarities and differences from computer operating systems in the traditional sense, such as Windows and Linux.

The role of the computer operating system is to encapsulate and manage the computer hardware; application software usually runs on the operating system and does not directly touch the hardware, regardless of the type of computer hardware product. Operating systems can also manage and schedule computer processes to enable multitasking. The application development interface provided by computer operating systems to developers can greatly improve the efficiency of software development, otherwise people have to write their own assembly programs.

The ROS operating system is a meta operating system, or a secondary operating system, for robots. ROS is not specifically responsible for computer process management and scheduling, but runs on the core of a computer operating system, encapsulates the robot hardware, provides a unified development interface for different robots and sensors, and interacts with information through a peer-to-peer communication mechanism that allows ROS provides an operating system-like process management mechanism and development interface, including hardware layer abstraction description, driver management, execution of common functions, passing messages between programs, and program release package management. It also provides a number of tool programs and libraries for acquiring, building, writing, and running programs for multi-machine integration.

2.1.4   Key Features of ROS

ROS has the following features.

1. Distributed architecture
     The main goal of ROS is to support code reuse in robotics research and development. The ROS system can be thought of as a framework of distributed processes, a mechanism that allows executable files to be individually designed and developed and loosely coupled at runtime. These processes can in turn be divided into project packages and project package sets, which can be easily shared and distributed. ROS also supports a federated system of code repositories, enabling collaboration to be distributed as well. This design (from the file system level to the community level) supports independent decisions about development and implementation, but also supports integration with the ROS base tools.
2. Multilingual neutrality

Because of the advantage that the distributed framework enables the executable files to be designed separately, different function node programs in ROS system can be implemented in different programming languages, which can avoid various technical problems such as long programming time, poor debugging effect, high syntax error rate and low execution efficiency caused by different programmers' preferences for programming languages and cultural differences. ROS supports a number of different languages, such as C++, Python and LISP, as well as different interface implementations of other languages. ROS uses a language-independent Interface Definition Language (IDL) and implements a wrapper around IDL for multiple programming languages, allowing transparent messaging.

3. Streamlining and integration

Robotics software development is moving toward modularity and systematization, and ROS encourages the development of all drivers and algorithms as separate libraries with no interdependencies independent of the ROS core system. The idea underlying ROS is to encapsulate complex code in separate libraries and additionally create small applications to show the specific functionality of the libraries, which allows the use of simple code beyond prototypes for porting and the modularity of ROS allows the code in each module to be compiled separately and is easily extended, making it particularly suitable for developing large distributed systems and large parallel processes. Code written in ROS can also be easily integrated with other robotic software frameworks, and ROS is currently available for integration with OpenRave, Orocos, and Player.

4. Rich toolkit

For better management of the ROS software framework, the ROS package also provides a number of gadget to facilitate developers in compiling and running various ROS components. These gadgets provide more flexible customization capabilities than building a large development and runtime environment. These gadgets can perform a variety of tasks, such as organizing and managing the source code structure, reading and configuring system parameters, graphically connecting end-to-end topologies, making measurements of band usage widths, depicting information data with images, automated generation of documentation, etc.

5. Open source and free

All source code for ROS is publicly released, with the development-language independent tools and the main client libraries (C++, Python, LISP) released under the BSD open-source license, which is free for commercial and research purposes. Other packages (such as those developed by developers themselves) have the flexibility to choose other open-source licenses such as Apache 2.0, GPL, MIT or even patent licenses. Users can determine for themselves whether a package has a license that meets their needs. This flexible open-source mechanism facilitates developers to debug and continuously correct errors at all levels of the ROS software, leading to continuous improvement of ROS.

2.2 Installation and Uninstallation of ROS

2.2.1 Versions of ROS

The specific version of Ubuntu should be selected based on the ROS version, each of which corresponds to one or two Ubuntu versions, as shown in Table 2.1. When installing, the ROS version should correspond to the system version, otherwise it will not be installed.

Ubuntu can be installed on a bare metal machine, or if you already have a Windows system, you can install a virtual machine and install Ubuntu on the virtual machine, or you can install Ubuntu dual system on top of Windows (recommended), the specific installation details can be searched for on the web according to the situation, so I won't go into details here.

Projects covered in this book are designed to run primarily on ROS Indigo, but some of the projects have also been tested on the Melodic version of ROS.

2.2.2 Installing and Configuring ROS Indigo

The ROS Indigo Deb package is only supported for installation on Ubuntu 13.10 (Saucy) and Ubuntu 14.04 (Trusty) systems. Here's how to install and configure Indigo on Ubuntu 14.04.

1. Configure the Ubuntu repository
   Configure the Ubuntu repository on your computer and set it up in the Software & Updates screen to allow three installation modes: restricted, universe, and multiverse (shown in Fig. 2.1).
2. Configure sources.list
   Add software sources to enable the system to install software from the packages.ros.org software source.

Table 2.1 ROS versions and corresponding ubuntu versions

Release date	ROS version	Corresponding to the Ubuntu version
May 2018	ROS Melodic	MoreniaUbuntu 18.04 (Bionic)/Ubuntu 17.10 (Artful)
May 2017	ROS Lunar	LoggerheadUbuntu 17.04 (Zesty)/Ubuntu 16.10 (Yakkety)/Ubuntu 16.04 (Xenial)
May 2016	ROS Kinetic	KameUbuntu 16.04 (Xenial)/Ubuntu 15.10 (Wily)
May 2015	ROS Jade	TurtleUbuntu 15.04 (Wily)/Ubuntu 14.04 (Trusty)
July 2014	ROS Indigo	IglooUbuntu 14.04 (Trusty)
...	...	...

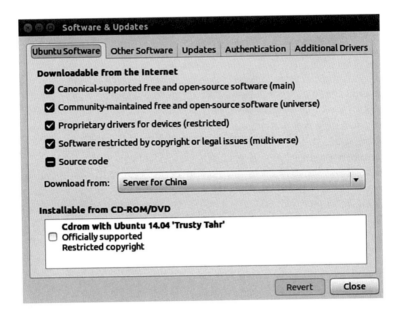

Fig. 2.1 Configuring the ubuntu software repository

```
$ sudo sh -c 'echo "deb http://packages.ros.org/ros/ubuntu
$(lsb_release -sc) main" > /etc/apt/sources.list.d/ros-latest.
list'
```

3. Configure the installation key (key)

   The command to configure the installation key is as follows.

```
$ sudo apt-key adv --keyserver hkp://pool.sks-keyservers.net --
recv-key 421C365BD9FF1F717815A3895523BAEEB01FA116
```

   If you have problems connecting to the keyserver, try replacing the server address in the above command with hkp://pgp.mit.edu:80 or hkp://keyserver. ubuntu.com:80.

4. Installation

   First, update the Debian package index with the following command.

```
$ sudo apt-get update
```

   If you are using Ubuntu 14.04, do not install the following software as it will prevent xserver from working properly.

```
$ sudo apt-get install xserver-xorg-dev-lts-utopic mesa-common-dev-
lts-utopic libxatracker-dev-lts-utopic libopenvg1-mesa-dev-lts-
utopic libgles2-mesa-dev-lts-utopic libgles1-mesa-dev-lts-utopic
```

```
libgl1-mesa-dev-lts-utopic libgbm-dev-lts-utopic libegl1-mesa-
dev-lts-utopic
```

Or try to fix the dependency by installing only the following tool.

```
$ sudo apt-get install libgl1-mesa-dev-lts-utopic
```

There are various libraries and tools in the ROS installation package, you can specify the installation method during installation, choose all or part of the package. ROS officially provides the following four default installation methods, developers have the flexibility to choose the package.

1. Officially recommended desktop full version installation: includes all official libraries, including ROS core package, rqt library, rviz, generic robot function library, 2D/3D emulator, navigation, and 2D/3D perception function package.
   The installation command is as follows.

```
$ sudo apt-get install ros-indigo-desktop-full
```

2. Desktop version installation: contains ROS core package, rqt library, rviz and generic robot function library, the installation command is as follows.

```
$ sudo apt-get install ros-indigo-desktop
```

3. Basic version installation: Includes the ROS core package, program building tools, and communication-related libraries, but no GUI tools. The installation command is as follows.

```
$ sudo apt-get install ros-indigo-ros-base
```

4. Individual package installation method: Install a specific ROS package individually (replace PACKAGE below with the package name).

```
$ sudo apt-get install ros-indigo-PACKAGE
```

For example.

```
$ sudo apt-get install ros-indigo-slam-gmapping
```

To find available packages, you can execute the following command.

```
$ apt-cache search ros-indigo
```

5. Initialize rosdep
   After installing the ROS system, you will need to initialize rosdep before you can start using it. rosdep is a tool that must be used for certain core ROS

functional components, and can be used to easily install some system dependencies for it when compiling certain source code.

```
$ sudo rosdep init
$ rosdep update
```

6. Environmental settings

In order to automatically configure the ROS environment variables each time you open a new terminal, you need to add the environment variables to the bash environment variable configuration file. Execute the following command.

```
$ echo "source /opt/ros/indigo/setup.bash" >> ~/.bashrc
$ source ~/.bashrc
```

If there are multiple versions of ROS installed, ~/.bashrc can only configure setup.bash for one currently used version.

If you only want to change the environment variables under the current terminal, you can execute the following command.

```
$ source /opt/ros/indigo/setup.bash
```

7. Install rosinstall

rosinstall is a standalone common command line tool in ROS that allows users to easily download many source trees to a particular ROS package with a single command.

To install this tool on Ubuntu, execute the following command.

```
$ sudo apt-get install python-rosinstall
```

2.2.3   Installing and Configuring ROS Melodic

Melodic is the latest version of the current ROS and is only supported on Ubuntu 18.04 (Bionic) and Ubuntu 17.10 (Artful). Here's how to install and configure Melodic on Ubuntu 18.04.

1. Configure the Ubuntu software repository.

The Ubuntu repositories are also configured in the "Software & Updates" interface and support three installation modes: restricted, universe and multiverse.
2. Add sources.list.

Configure the repository so that it can install software from packages.ros.org.

```
$ sudo sh -c 'echo "deb http://packages.ros.org/ros/ubuntu
$(lsb_release -sc) main" > /etc/apt/sources.list.d/ros-latest.
list'
```

3. Add the key.

```
$ sudo apt-key adv --keyserver hkp://ha.pool.sks-keyservers.net:80
--recv-key 421C365BD9FF1F717815A3895523BAEEB01FA116
```

4. Installation.

First, update the Debian package index to the latest version with the following command.

```
$ sudo apt update
```

If installing the full desktop version, execute the following command.

```
$ sudo apt install ros-melodic-desktop-full
```

If installing the desktop version only, execute the following command.

```
$ sudo apt install ros-melodic-desktop
```

If installing only the base version, execute the following command.

```
$ sudo apt install ros-melodic-ros-base
```

If doing a single package installation, execute the following command.

```
$ sudo apt install ros-melodic-PACKAGE
```

For example.

```
$ sudo apt install ros-melodic-slam-gmapping
```

You can find available packages by executing the following command.

```
$ apt search ros-melodic
```

5. Initialize rosdep.

Before you can start using ROS, you also need to initialize rosdep. rosdep can be used to install some system dependencies for certain source code when it needs to be compiled, and it is also a necessary tool for some core ROS functional components.

```
$ sudo rosdep init
$ rosdep update
```

6. Environmental settings.
   Add the ROS environment variable to the bash session as follows.

```
$ echo "source /opt/ros/melodic/setup.bash" >> ~/.bashrc
$ source ~/.bashrc
```

If you only want to change the environment variables under the current terminal, you can execute the following command.

```
$ source /opt/ros/melodic/setup.bash
```

7. Build package dependencies.
   So far, we have installed the components needed to run the core ROS packages. Often, developers will also need to create and manage their own ROS workspace, which involves installing various toolkits separately, depending on the requirements. For example, to install the rosinstall tool and other dependencies used to build ROS packages, execute the following command.
   $ sudo apt install python-rosinstall python-rosinstall-generator python-wstool build-essential

2.2.4 Uninstallation of ROS

If the Indigo version of ROS was installed with apt-get, it can be uninstalled with the following command.

```
$ sudo apt-get remove ros-<ROS name>-*
```

Where inside < > is the ROS version name, e.g. indigo. if the uninstall is successful, the indigo directory in the ROS folder in the /opt directory will be deleted.

2.3 Resources for Further Learning

Here are some resources for learning ROS and related knowledge, interested readers can refer to these materials for in-depth learning.

ROS Wikipedia official English tutorial: http://wiki.ros.org/
ROS Wikipedia official Chinese tutorial: http://wiki.ros.org/cn
Computer Vision Related Resources.
OpenCV: http://opencv.org/
PCL: http://pointclouds.org/

Openni: http://www.openni.org/
Reconstructme: http://reconstructme.net/
PrimeSense: http://www.primesense.com/
OpenKinect: http://openkinect.org/wiki/Main_Page
Related Learning Sites.
CreateSmart: https://www.ncnynl.com/

Exercises

1. What are the benefits of using ROS?
2. What is the difference between ROS and a computer operating system in the traditional sense?
3. How to confirm that your ROS has been installed successfully?

Further Reading

1. Song, Huixin, Shao, Zhenzhou. New developments in robot operating systems [J]. Automation Expo, 2016(9):32–33.
2. Thunderbird.com. The state of the art and future optimization of robot operating systems [EB/OL]. https://www.leiphone.com/news/201612/PpjEsSWU6RwN1yVI.html.
3. Wikipedia. ROS/Introduction [EB/OL]. http://wiki.ros.org/cn/ROS/Introduction.
4. Blogland. Robot Operating System ROS [EB/OL]. https://www.cnblogs.com/qqfly/p/58513-82.html.
5. Zhou X.S., Yang G., Wang L., et al. Robot operating system ROS principles and applications [M]. Beijing: machinery industry publishing house, 2017.
6. Wikipedia. ROS/Distributions [EB/OL]. http://wiki.ros.org/Distributions.

Chapter 3
The Framework and Fundamental Use of ROS

After the correct installation of ROS, we will continue to learn the ROS framework with the initial usage. In the ROS framework section, we will help readers to understand the composition and resource distribution of ROS fully. The ROS framework consists of three levels: file system level, computational graph level, and community level; after understanding the basic framework of ROS, we will introduce the basics of ROS usage, mainly including the introduction of catkin, workspaces and their creation, creation and compilation of project packages, creation and compilation of ROS nodes and running them, the use of roslaunch, the creation of ROS messages and services, how to write test message publishers and subscribers in C++ or Python, how to write test Servers and Clients in C++ or Python, and more. The content of this chapter is the basis for conducting subsequent chapters, and you should be proficient in it.

3.1 ROS Framework

To facilitate developers' understanding, the ROS system framework can generally be divided into three levels: the file system level, the computational graph level, and the community level, with the roles of each level shown in Fig. 3.1.

3.1.1 File System Level

The file system level refers primarily to the ROS directories and files that are visible on the hard disk, including.

1. Project Package (Package): It is the main unit for organizing software in the ROS. A project package may contain ROS runtime nodes, a dependency library,

© The Author(s), under exclusive license to Springer Nature Singapore Pte Ltd. 2023
F. Duan et al., *Intelligent Robot*, https://doi.org/10.1007/978-981-19-8253-8_3

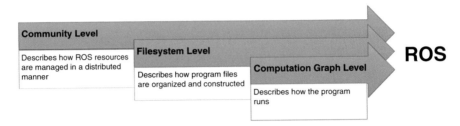

Fig. 3.1 ROS framework hierarchy

datasets required for operation, node configuration files, or other valuable files. The project package is a compilation item, and a release item at the atomic-level in ROS, i.e., the smallest item compiled and released is the project package.

2. Metapackage: Metapackages are specialized project packages that are used to represent only a set of related project packages.
3. Package Manifest: The manifest (package.xml) provides metadata related to the project package, including the package name, version, description, license information, dependencies, and other metadata information, such as the exported project package.
4. Repository: Repository is a collection of packages that share a standard VCS system. Packages that share the same VCS system have the same version and can be published together with the catkin automated publishing tool bloom. Of course, a repository can also contain only one package.
5. Message (msg) type: The message type defines and describes the data structure of ROS messages. The message type is stored in my_package/msg/mymessagetype. msg and is used to define the data structure of the messages sent in ROS.
6. Service (srv) type: Service type describes the data structure of ROS service. Service types are stored in my_package/srv/myservicetype.srv, which defines the data structure of the request and service response in ROS.

3.1.2 Calculation of Graph Levels

The computational graph is a peer-to-peer network of ROS processes in which ROS integrates and processes data. The basic concepts of the ROS computational graph include Node, Master, Parameter Server, Message, Service, Topic, and Bag, which all provide data to the computational graph in different ways.

1. Nodes: Nodes are processes that perform operations. A robot control system usually contains multiple nodes. For example, a node that controls a laser rangefinder, a node that controls a wheel motor, a node that performs positioning, a node that performs path planning, a node that provides a graphical view of the system, etc. ROS nodes are typically written using the ROS client library (roscpp or rospy).

Fig. 3.2 Nodes communicate directly with the topic for message communication

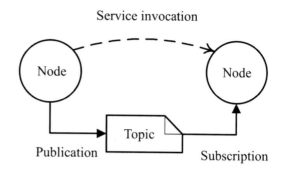

2. Master Controller: The ROS master controller provides name registration and lookup services for the entire computational graph. Through the master controller, the nodes can find each other by name, exchange messages or invoke services.
3. Parameter Server: The parameter server allows data keys to be stored in a central location. It is currently part of the main controller.
4. Messages: Nodes communicate with each other by passing messages. A message is simply a data structure that supports standard data types (integer, float, boolean, etc.) and the corresponding arrays. Messages can contain arbitrarily nested structures and arrays (similar to the structure in C).
5. Topics: Messages are delivered through a transport system with publish/subscribe semantics. Nodes implement message delivery by publishing messages to a given topic. The topic corresponds to a defined name that identifies the content of the message, and nodes interested in a certain type of data can obtain messages by subscribing to the corresponding topic. A topic can have multiple publishers and subscribers publishing and subscribing to messages simultaneously, and a node can also publish and/or subscribe to multiple topics at the same time. Strictly speaking, publishers and subscribers do not need to be aware of each other's existence, and the idea is to disconnect the producer from the user of the information. Logically, topics can be thought of as strongly typed message buses. Each bus has a name, and anyone can connect to the bus to send or receive messages, as long as they are of the correct type.
6. Services: The publish/subscribe model is a flexible communication model, but its many-to-many one-way transmission is unsuitable for distributed systems. Because distributed systems often require request/response interactions, so ROS provides services as a communication method. A service consists of a pair of message structures: a request message and an answer message.
7. Packets: Packets are structures used to store and playback ROS message data, and they are an important mechanism for preserving runtime data (e.g., sensor data). This data can be difficult to collect but is necessary for developing and testing algorithms.

The message communication process between the node and the topic is shown in Fig. 3.2.

In the ROS computational graph, the ROS master controller acts as a name service, and it keeps and maintains registration information for all topics and services currently running in the ROS. Each node communicates with the master controller and registers its registration information when it starts. When these nodes communicate with the master controller, they can receive information from other registered nodes and make the appropriate connections. The master controller will also call back to these nodes when the registration information changes, which allows the nodes to create connections when running new nodes dynamically.

This architecture supports decoupled operations, where names are the primary means of building larger, more complex systems. Names have a significant role in ROS, and all of nodes, topics, services, and parameters must have definite names. Each ROS client library can remap names so that compiled programs can be reconfigured at runtime and run in a different computational graph topology.

For example, to drive a Hokuyo laser rangefinder, you can start a driver with the name hokuyo_node that talks to the laser rangefinder and creates a scan topic that posts sensor_msgs/LaserScan messages to the topic. In order to process the scan data from the laser rangefinder, we can write a filter node with the name laser_filters that subscribes to the messages of the scan topic. After subscribing, our filter node will automatically start receiving messages from the laser rangefinder.

So how are these two nodes decoupled? All the hokuyo_node node does is publish scans, and it doesn't know if any nodes subscribe to them. All the laser_filters filter node does is subscribe to scans, and it doesn't know if any nodes publish scans either. These two nodes can be started, terminated and restarted individually at will without raising any errors.

3.1.3 Community Level

ROS community level refers to ROS resources used to enable different communities to exchange software and knowledge. These resources include.

1. Distribution: A ROS distribution is a collection of installable, versioned ROS project packages. See http://wiki.ros.org/Distributions.
2. Repository: Repository is a federated network of ROS- dependent code repositories where different organizations can develop and distribute their own software. See http://wiki.ros.org/Repositories.
3. ROS Wiki Community: The ROS Wiki Community is the main forum for maintaining information related to ROS. Anyone can register for a community account, upload their own documentation, post corrections or updates, provide tutorials, etc. See http://wiki.ros.org/Documentation.

Other resources include.

1. Bug Ticket System: See http://wiki.ros.org/Tickets for information on file tickets.
2. Mailing List: The ROS user mailing list provides the main communication channel for ROS updates and is the most active forum for ROS software issues. See http://wiki.ros.org/action/show/Su-pport?action=show&redirect=Mailing +Lists.
3. ROS Answer: It is a Q&A site that answers ROS-related questions. See https:// answers.ros.org/questions/.
4. Blog: Blog provides regular updates, including photos and videos. See http:// www.ros.org/news/.

3.2 Basics of ROS Use

Before we can officially use ROS, we need to have a preliminary understanding of the ROS file system, the creation of project packages, and how they work. This is the foundation for developing ROS and helps us to understand better how ROS works.

3.2.1 Overview of Catkin

catkin is the official compilation system for ROS and the successor to Rosbuild, which is the original compilation system for ROS. catkin combines CMake macro commands and Python scripts to provide some functionality over the normal CMake workflow. catkin is designed to be more practical than Rosbuild, supporting the better distribution of project packages, providing better cross-compilation support and portability. catkin's workflow is similar to that of CMake, and catkin project packages can be built as a standalone project, just like a normal CMake project. catkin also provides the concept of workspace, which allows multiple interdependent packages to be built at the same time.

The name catkin comes from the tail-like flaps on the willow tree—indicating that Willow Garage created the catkin. For more information on the catkin, see http:// wiki.ros.org/catkin/conceptual_overview.

3.2.2 Workspaces and Their Creation Methods

1. Introduction to catkin workspace
The catkin workspace is the folder where catkin project packages are modified, built, and installed. The following is a recommended layout for a typical catkin workspace.

```
workspace_folder/ -- workspace
  src/ -- source space
   CMakeLists.txt -- the "top-level" CMake file
   package_1/
    CMakeLists.txt
    package.xml
    . . .
   package_n/
    CATKIN_IGNORE -- exclude package_n from the executing project
package, optionally with an empty file
    CMakeLists.txt
    package.xml
    . . .
  build/ -- compile space
   CATKIN_IGNORE -- prevent catkin from browsing this directory
  devel/ - development space (set by CATKIN_DEVEL_PREFIX)
   bin/
   etc/
   include/
   lib/
   share/
   .catkin
   env.bash
   setup.bash
   setup.sh
   . . .
  install/ -- installation space (set by CMAKE_INSTALL_PREFIX)
   bin/
   etc/
   include/
   lib/
   share/
   .catkin
   env.bash
   setup.bash
   setup.sh
   . . .
```

A catkin workspace contains up to four different spaces, each playing a different role in the software development process.

(a) Source Space (src): The source space contains the source code of the catkin project package. Here, the user can extract/check/copy the source code of the project package to be generated. Each folder in the source space contains one or more catkin project packages. The space can remain unchanged through operations such as configuration, compilation, or installation. The root directory of the source space contains a symbolic links to the CMakeLists.txt file, which is at the top level of the catkin. This file is called by CMake during the configuration of a catkin project in the workspace. It can be created by calling catkin_init_workspace in the source space directory.

(b) Build Space (build): This is where the project packages in source space are compiled when CMake is invoked and is used to store cache information and other intermediate files generated during the workspace compilation. The build space does not have to be contained in the workspace, nor does it have to be located outside of the source space (but this is usually recommended).

(c) Development Space (devel): The development space is used to hold the target programs that have been compiled prior to the installation of the project package. The target program is organized in the development space in the same way as its installation layout, which provides a useful testing and development environment without the need to invoke the installation steps. The location of the development space is controlled by a catkin special CMake variable CATKIN_DEVEL_PREFIX, which is located by default in <build space>/develspace. This is the default behavior since if a CMake user calls cmake from within the build folder, it will modify the current directory outside of the contents, which may confuse the user. It is recommended that the development space directory be set as a counterpart to the build space directory.

(d) Install Space (install): Once the targets have been compiled, they can be installed into the install space by invoking the install target (usually using make install). The install space does not have to be included in the workspace. Since the install space is set by CMAKE_INSTALL_PREFIX, it is located by default in /usr/local directory and should not be used (as uninstallation is nearly impossible and using multiple ROS distributions does not work).

(e) Result Space: The generic term "result space" is used when referring to a folder in the development space or installation space.

2. Create catkin workspace

When ROS is installed, catkin is included by default. Next we start creating and compiling a catkin workspace.

```
$ mkdir -p ~/robook_ws/src # robook_ws is the workspace created, src
is the source space, which contains the source code of the project
package
$ cd ~/robook_ws/
$ catkin_make
```

The catkin_make command is a convenient tool when compiling ROS packages, and it is described in detail at http://wiki.ros.org/catkin/Tutorials/using_a_workspace. After running catkin_make for the first time, it will create a cmakelists.txt link in the src folder and also generates a build and devel folder in the current directory. There are several setup.*sh files in the devel folder. Using the source command on any of these files will set the current workspace at the top level of the ROS working environment.

The newly generated setup.sh file can be processed with the following source command.

```
$ source devel/setup.bash
```

To ensure that the setup script has correctly overwritten the workspace, ensure that the ROS_PACKAGE_PATH environment variable contains the workspace directory, which can be viewed using the following command.

```
$ echo $ROS_PACKAGE_PATH
```

The output contains the following.

```
/home/youruser/robook_ws/src:/opt/ros/melodic/share
```

Where youruser is the user name and melodic is the ROS name.

Here, we add the created ros workspace to the global path so that we don't need to use the source command for the setup.sh file under the project package every time we open a terminal command-line.

```
$ echo "source robook_ws/devel/setup.bash" >> ~/.bashrc
$ source ~/.bashrc
```

3.2.3 Creating ROS Project Packages

We will use the catkin_create_pkg command to create a new catkin project package. catkin_create_pkg command requires package_name (the name of the project package) and can be followed by some other project packages (depend) that need to be depended on if needed.

```
$ catkin_create_pkg <package_name> [depend1] [depend2] [depend3]
```

Next, we can start creating our own project package. First, switch to the src directory of the robook_ws workspace created earlier with the following command.

```
$ cd ~/robook_ws/src
```

Create a project package named "ch3_pkg" with the following command.

```
# This project package depends on std_msgs, roscpp, rospy
$ catkin_create_pkg ch3_pkg std_msgs rospy roscpp
```

As you can see, a folder named "ch3_pkg" is created in the robook_ws/src directory. This folder contains a package.xml file and a CMakeLists.txt file, both of which automatically contains some of the information provided during the execution of the catkin_create_pkg command.

3.2.4 Compiling the ROS Project Package

First, use the source command for the environment configuration (setup) file.

```
$ source /opt/ros/melodic/setup.bash
```

Next, the project package is compiled using the command-line tool catkin_make. catkin_make simplifies the standard workflow of catkin and its work can be thought of as calling cmake and make in sequence within the standard workflow of CMake. Use the following.

```
# Under the catkin workspace
$ catkin_make [make_targets] [-DCMAKE_VARIABLES=...]
Now, switch to the workspace that has been created and use catkin_make to
compile.
$ cd ~/robook_ws/
$ catkin_make
```

3.2.5 Creating ROS Nodes

We will use the creation of a hello.cpp node as an example to describe the process of creating a ROS node. We will write a node that outputs the string "Hello, world!" and run it.

1. Go to the src directory in the "ch3_pkg" project package that contains the source code, create the hello.cpp file, and then use the gedit editor to edit it.

   ```
   $ cd ~/robook_ws/src/ch3_pkg/src
   $ touch hello.cpp
   $ gedit hello.cpp
   ```

2. Change the following code in hello.cpp.

   ```
   #include <ros/ros.h>
   int main(int argc, char **argv) {
   }
   ```

3. Modify the CMakeLists.txt file to include information about the new node being written.

```
cmake_minimum_required(VERSION 2.8.3)
project(ch3_pkg)
```

```
find_package(catkin REQUIRED COMPONENTS
```

```
add_executable(hello ./src/hello.cpp)
```

```
target_link_libraries(hello ${catkin_LIBRARIES})
```

3.2.6 Compiling and Running ROS Nodes

In this section, we are going to compile and run the nodes created in the previous section.

1. Compile the program in the ~/robook_ws directory.

```
$ catkin_make
```

2. After successful compilation, run the node. In a new terminal, run the following command.

```
$ roscore
```

3. Open a new terminal, go to the robook_ws directory, and run the following command.

```
$ rosrun ch3_pkg hello
```

   If the workspace has not been added to the global path before, rosrun will not find the ch3_pkg project package. At this point, we need to execute the source command in the robook_ws directory before execute the rosrun command.

```
$ source devel/setup.bash
$ rosrun ch3_pkg hello
```

   As you can see, there is an output of "Hello,world!", as follows.

```
[ INFO] [1548582101.184988498]: Hello, world!
```

3.2.7 Use of Roslaunch

roslaunch is a command that runs one or more nodes in a given project package or sets execution options. The syntax is as follows.

```
$ roslaunch [package] [filename.launch]
```

Let's run two little turtle turtlesim as an example of how it works.

1. Switch to the ch3_pkg project package directory that we have created earlier.

```
$ roscd ch3_pkg
```

2. Create a launch folder.

```
$ mkdir launch
$ cd launch
```

3. Create a launch file, name it example_launch.launch, and modify its contents as follows.

```
<launch>

    <group ns="turtlesim1">
        <node pkg="turtlesim" name="sim" type="turtlesim_node"/>
    </group>

    <group ns="turtlesim2">
        <node pkg="turtlesim" name="sim" type="turtlesim_node"/>
    </group>

    <node pkg="turtlesim" name="mimic" type="mimic">
        <remap from="input" to="turtlesim1/turtle1"/>
        <remap from="output" to="turtlesim2/turtle1"/>
    </node>

</launch>
```

Several of the tags used in the document are described below.

<launch>: It describes the tags required to run the node using the roslaunch command.

<group>: It is a tag for grouping the specified nodes. Where the ns option refers to the namespace, which is the name of the group.

<node>: It describes the node on which roslaunch is running, with options pkg (the name of the project package), type (the name of the node that is actually

running), and name (the name of the node corresponding to type when it is running, which can be the same as or different from the name of type).

4. Parsing the launch file paragraph by paragraph.

```
</launch>
```

The purpose of the opening launch tag is to declare that this is a launch file.

```
<group ns="turtlesim1">
    <node pkg="turtlesim" name="sim" type="turtlesim_node"/>
</group>

<group ns="turtlesim2">
    <node pkg="turtlesim" name="sim" type="turtlesim_node"/>
</group>
```

Create two groups of nodes named turtlesim1 and turtlesim2. Both groups use the same node with the name sim so that two turtlesim simulators can be started at the same time without naming conflicts.

```
<node pkg="turtlesim" name="mimic" type="mimic">
      <remap from="input" to="turtlesim1/turtle1"/>
      <remap from="output" to="turtlesim2/turtle1"/>
</node>
```

Start the mimic node. Rename the inputs and outputs of all topics as turtlesim1 and turtlesim2 respectively, and making turtlesim2 mimic turtlesim1.

```
</launch>
```

The closing launch tag is the marker for the end of the launch file.

5. The launch file can then be started with the roslaunch command.

```
$ roslaunch ch3_pkg example_launch.launch
```

At this point, you can see that two turtlesim have been started, as shown in Fig. 3.3.

You can also use the rostopic command to send a speed setting message to make both turtles move at the same time. Enter the following command in a new terminal.

```
$ rostopic pub /turtlesim1/turtle1/cmd_vel geometry_msgs/Twist -r
1 -- '[2.0, 0.0, 0.0]' '[0.0, 0.0, -1.8]'
```

Two small turtles can be seen moving at the same time, as shown in Fig. 3.4.

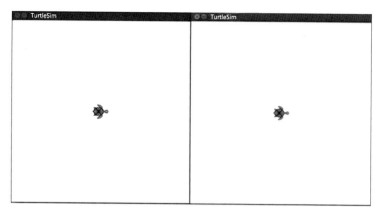

Fig. 3.3 turtlesim1 and turtlesim2 start

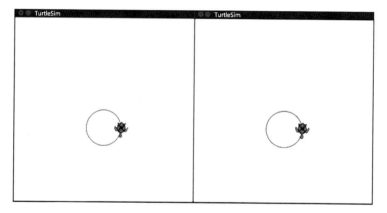

Fig. 3.4 Simultaneous motion of turtlesim1 and turtlesim2

3.2.8   Creating ROS Messages and Services

First, let's understand what messages and services are. A message is a piece of information sent from one process (node) to other processes (nodes) in ROS, and nodes complete communication with each other through messages. The data structure of a message is called a message type, and the ROS system provides many standard types of messages that users can use directly. If they want to use some non-standard types of messages, they need to define that type of messages themselves.

Service is a request/response communication process between processes (nodes) in ROS. There are special occasions when nodes need to communicate with each other in an efficient peer-to-peer manner and get a timely response, which requires the use of services for interaction. The node that provides the service is called the server, and the node that initiates a request to the server and waits for a response is called the client. The client initiates a request and gets a response from the server so that a service communication process is completed. The data structure of the service request/response is called the service type. The definition of service type borrows from the way message type is defined, the difference between them is that message data is the information passed during the many-to-many broadcast communication between ROS processes (nodes), and service data is the information passed during the point-to-point request/response communication between ROS processes (nodes).

Messages in ROS are defined by the msg file. This file is a simple text describing the types of messages used in ROS and can be used to generate source code in different languages. It is stored in the msg directory of the project package.

Each line in the msg file declares a data type and variable name, and the following data types can be used.

```
int8, int16, int32, int64 (plus uint*)
float32, float64
string
time, duration
Other msg files
variable-length array and fixed-length array
```

ROS also has a special data type: Header, which contains a timestamp and coordinate system information.

The following is an example of a msg file.

```
Header header
string child_frame_id
geometry_msgs/PoseWithCovariance pose
geometry_msgs/TwistWithCovariance twist
```

In ROS, an srv file describes a service and contains two parts, "request" and "response", separated by "---". srv files are stored in the srv file is stored in the srv directory of the project package.

The following is an example of an srv file.

```
int64 A
int64 B
--
int64 Sum
# A, B is the request, Sum is the response
```

The steps to create a message (msg) in the ch3_pkg project package are as follows.

1. Create a msg folder in the project package, create the Num.msg file in it, and add a line declaring.

```
$ cd ~/robook_ws/src/ch3_pkg
$ mkdir msg
$ echo "int64 num" > msg/Num.msg
```

2. To ensure that the msg file is supported by C++, Python, or other languages, the package.xml file needs to be modified to include the following two statements.

```
<build_depend>message_generation</build_depend>
<exec_depend>message_runtime</exec_depend>
```

3. Modify the CMakeLists.txt file. In the find_package function, add a dependency on message_generation.

```
find_package(catkin REQUIRED COMPONENTS
 roscpp
 rospy
 std_msgs
 message_generation
 )
```

4. Set operational dependencies.

```
catkin_package(
 CATKIN_DEPENDS message_runtime
 )
```

5. Modify the following code block.

```
add_message_files(
 FILES
 Num.msg
 )
```

6. Add the generate_message() function.

```
generate_messages(
 DEPENDENCIES
 std_msgs
 )
```

At this point, the CMakeLists.txt file is modified, and the job of creating the message is done.

After creating a message, you can check if the message can be recognized by ROS using the rosmsg command-line tool.

```
$ rosmsg show ch3_pkg/Num
```

The output is.

```
int64 num
```

The steps to create a service (srv) in the ch3_pkg project package are as follows.

1. Create a srv folder in the project package.

```
$ roscd ch3_pkg
$ mkdir srv
```

2. Services can be created manually, or copied from other project packages. Here, we use the service files from the rospy_tutorials project package.

```
# roscp: copy files from project package
$ roscp rospy_tutorials AddTwoInts.srv srv/AddTwoInts.srv
```

3. Similarly, the CMakeLists.txt file needs to be modified.
   First, CMakeLists.txt is going to add a dependency on message_generation using the find_package() function (this example was added when the msg file was created).
4. Modify the following code block.

```
add_service_files(
  FILES
  AddTwoInts.srv
)
```

This completes the work of creating a service.
You can check if ROS recognizes the service by using the rossrv command-line tool. The command is as follows.

```
$ rossrv show ch3_pkg/AddTwoInts
The output is as follows.
int64 a
int64 b
--
int64 sum
```

If you need to use msg and srv, you also need to perform the following steps.

1. In CMakeLists.txt, remove the comments from the generate_messages() function.

```
# This example depends on std_msgs. No need to add roscpp, rospy
generate_messages(
 DEPENDENCIES
 std_msgs
 )
```

2. Recompile the project package.

```
$ cd ~/robook_ws
$ catkin_make
```

3.2.9   Writing a Simple Message Publisher and Subscriber (C++ Implementation)

In this section, we use C++ to write a simple message publisher and subscriber.

```
First, create the publisher node (talker), whose function is to
continuously broadcast messages across the ROS network. Switch to the
previously created ch3_pkg/src path and create the talker.cpp file and
edit it using gedit.
$ cd ~/robook_ws/src/ch3_pkg/src
$ touch talker.cpp
$ gedit talker.cpp
Use the following code.
#include "ros/ros.h" // reference most of the common header files in ROS
#include "std_msgs/String.h" /* header file automatically generated by
String.msg file
std_msgs/String messages are stored in the std_msgs package */
#include <sstream>

int main(int argc, char **argv)
{
  ros::init(argc, argv, "talker"); /* Initialize the ROS.
Name remapping via the command-line,
It is also possible to specify the name of the node
The name of the node must be unique during the operation. */

  ros::NodeHandle n; // create handle for process node

  ros::Publisher chatter_pub = n.advertise<std_msgs::String>
("chatter", 1000);
/*
NodeHandle::advertise(): returns a ros:Publisher object.
First parameter chatter: post a message of type std_msgs/String on
```

chatter (topic name), whereupon master (node manager) will tell all
nodes subscribed to the chatter topic that it is about to be published by
data.
Second parameter: is the size of the release sequence.
```
*/
```

```
  ros::Rate loop_rate(10); //Specify the self-loop frequency. This
sentence runs at a frequency of 10Hz

  int count = 0;
  while (ros::ok()) // if ros::ok() returns false, all ROS calls will
fail
  {
      std_msgs::String msg;
      std::stringstream ss;
      ss << "hello world " << count;
      msg.data = ss.str();
/*
```
A "message adaptive" class generated from a msg file is used to broadcast
messages in the ROS network. It has only one data member, "data".
```
*/

      ROS_INFO("%s", msg.data.c_str()); //can replace printf/cout etc

      chatter_pub.publish(msg); // send a message to all nodes subscribed
to the chatter topic

      ros::spinOnce();

      loop_rate.sleep(); // call ros::Rate object to sleep for a period of
time to make the posting frequency/10Hz
      ++count;
  }

  return 0;
}
```
Next, write the subscriber node (listener). Create the listener.cpp file
under the ch3_pkg/src path and edit it using gedit.
```
$ touch listener.cpp
$ gedit listener.cpp
```
Use the following code.
```
#include "ros/ros.h"
#include "std_msgs/String.h"

void chatterCallback(const std_msgs::String::ConstPtr& msg)
// Callback function will be called when the chatter topic is received.
{
  ROS_INFO("I heard: [%s]", msg->data.c_str());
}
```

```
int main(int argc, char **argv)
{
  ros::init(argc, argv, "listener");

  ros::NodeHandle n;

  ros::Subscriber sub = n.subscribe("chatter", 1000,
chatterCallback);
/* NodeHandle::subscribe(): returns the ros::Subscriber object. It
must be active until it is no longer subscribed to the message. When this
object is destroyed, it will automatically unsubscribe from the chatter
topic's messages.
First parameter "chatter": tells master (node manager) that it will
subscribe to messages on the chatter topic. The chatterCallback()
function will be called by ROS when a message is posted.
Second parameter: queue size.
*/

  ros::spin(); // enter self-loop

  return 0;
}
```

Finally, compile the node. First, you need to modify the CMakeLists.txt
file by adding the following statement to the end of the file.

```
include_directories(include ${catkin_INCLUDE_DIRS})

add_executable(talker src/talker.cpp)
target_link_libraries(talker ${catkin_LIBRARIES})

add_executable(listener src/listener.cpp)
target_link_libraries(listener ${catkin_LIBRARIES})
```

Then, go back to the workspace path for compilation.

```
$ cd ~/robook_ws
$ catkin_make
```

3.2.10   Writing a Simple Message Publisher and Subscriber (Python Implementation)

In this section, we use the Python language to write a simple message publisher and subscriber.

First, write the publisher node (talker). In the ch3_pkg project
package, create the scripts directory to hold the Python code. Then
create the talker.py file and edit it using gedit as follows.

```
$ roscd ch3_pkg
$ mkdir scripts
$ cd scripts
$ touch talker.py
$ gedit talker.py
```

Use the following code.

```python
#! /usr/bin/env python
# license removed for brevity
# Ensure that the script is a script executed using Python.

import rospy
from std_msgs.msg import String
# Import the rospy client library and std_msgs.msg to reuse the std_msgs/
String message type.

def talker():
    # Define the talker interface
    pub = rospy.Publisher('chatter', String, queue_size=10)
    # Nodes post chatter topics, using String character types, with a queue
size of 10.
    rospy.init_node('talker', anonymous=True)
    # Initialize nodes

    rate = rospy.Rate(10)
    # Create a Rate object to control the frequency of posting topic
messages, in this case 10Hz

    while not rospy.is_shutdown():
        # Quit if it returns false, or keep running if there is no return value
        hello_str = "hello world %s" % rospy.get_time()
        rospy.loginfo(hello_str)
        # Output debug information on the screen, and write both node log file
and rosout node simultaneously
        pub.publish(hello_str)
        # Post String messages in chatter topics
        rate.sleep()
        # Combine with rospy.Rate() to keep message delivery frequency

if __name__ == '__main__':
    try:
        talker()
    except rospy.ROSInterruptException:
        pass
```

Modify permissions to be executable.

```
$ chmod +x talker.py
```

Next, write the subscriber node (listener). Create the listener.py file in the scripts directory.

```
$ touch listener.py
$ gedit listener.py
```

Use the following code.

```python
#! /usr/bin/env python
import rospy
from std_msgs.msg import String

def callback(data):
    rospy.loginfo(rospy.get_caller_id() + "I heard %s", data.data)
```

```
def listener():
  rospy.init_node('listener', anonymous=True)
  '''
  anonymous=True will tell rospy to generate a unique node name that
  Thus allowing multiple listener.py to run simultaneously
  (ROS requires each node to have a unique name, and if have the same name,
  it will abort the previous node with the same name)
  '''

  rospy.Subscriber("chatter", String, callback)
  '''
  The node subscribes to the topic chatter, and the message type is
  std_msgs.msgs.String.
  Once a new message is received, the callback function is triggered to
  process this information and take the message as the first argument passes
  to the function.
  '''

  rospy.spin() # Keeps the node running until the program is closed.

if __name__ == '__main__':
  listener()
```

Modify permissions to be executable.

```
$ chmod +x listener.py
```

Finally, compile the node. Enter the workspace and run the following command.

```
$ cd ~/robook_ws
$ catkin_make
```

3.2.11 Testing Message Publisher and Subscriber

In this section, let's look at how to test message publishers and subscribers.

First, start the publisher and run the talker node you created earlier:

```
$ roscore
$ cd ~/robook_ws
$ source ./devel/setup.bash
$ rosrun ch3_pkg talker (C++)
$ rosrun ch3_pkg talker.py (Python)
```

The output message can be seen as follows.

```
......
[INFO] [1548619061.279528]: hello world 1548619061.28
[INFO] [1548619061.379443]: hello world 1548619061.38
[INFO] [1548619061.479474]: hello world 1548619061.48
[INFO] [1548619061.579381]: hello world 1548619061.58
[INFO] [1548619061.679409]: hello world 1548619061.68
......
```

Then, start the subscriber and run the listener subscriber node created earlier:

```
$ rosrun ch3_pkg listener (C++)
$ rosrun ch3_pkg listener.py (Python)
The output message can be seen as follows.
......
[INFO] [1548619061.281000]: /listener_23804_1548619061046I heard
hello world 1548619061.28
[INFO] [1548619061.381006]: /listener_23804_1548619061046I heard
hello world 1548619061.38
[INFO] [1548619061.481059]: /listener_23804_1548619061046I heard
hello world 1548619061.48
[INFO] [1548619061.580614]: /listener_23804_1548619061046I heard
hello world 1548619061.58
[INFO] [1548619061.680674]: /listener_23804_1548619061046I heard
hello world 1548619061.68
......
```

3.2.12 Writing a Simple Server and Client (C++ implementation)

In this section, we will learn how to write Server and Client nodes in C++.

```
First, write the Server node, whose role is to take two integer numbers
and return their sum. This will use the srv file created earlier.
Create the example_server.cpp file in the ch3_pkg/src path.
$ cd ~/robook_ws/src/ch3_pkg/src
$ touch example_server.cpp
$ gedit example_server.cpp
Use the following code.
#include "ros/ros.h"
#include "ch3_pkg/AddTwoInts.h"

bool add(ch3_pkg::AddTwoInts::Request &req,
    ch3_pkg::AddTwoInts::Response &res)
/*
Provides a service for summing two int values: the int value is fetched
from the request, the return data is filled in the response, and the data
type is defined inside the srv file. The function returns a boolean value.
*/
{
  res.sum = req.a + req.b;
  ROS_INFO("request: x=%ld, y=%ld", (long int)req.a, (long int)req.b);
  ROS_INFO("sending back response: [%ld]", (long int)res.sum);
  return true;
}
/*
The two values are added together and stored in response. Record request
and response information.
Completes the calculation and returns true
*/
```

```cpp
int main(int argc, char **argv)
{
  ros::init(argc, argv, "add_two_ints_server");
  ros::NodeHandle n;

  ros::ServiceServer service = n.advertiseService("add_two_ints",
  add);
  // Create the service and publish it within the ROS

  ROS_INFO("Ready to add two ints.");
  ros::spin();

  return 0;
}
```

Next, write the Client node, creating the example_client.cpp file in the ch3_pkg/src path.

```
$ touch example_client.cpp
$ gedit example_client.cpp
```

Use the following code.

```cpp
#include "ros/ros.h"
#include "ch3_pkg/AddTwoInts.h"
#include <cstdlib>

int main(int argc, char **argv)
{
  ros::init(argc, argv, "add_two_ints_client");
  if (argc != 3)
  {
    ROS_INFO("usage: add_two_ints_client X Y");
    return 1;
  }

  ros::NodeHandle n;
  ros::ServiceClient client = n.serviceClient<ch3_pkg::AddTwoInts>
  ("add_two_ints");
  // Create client, ros::ServiceClient object for calling service
  ch3_pkg::AddTwoInts srv;
  srv.request.a = atoll(argv[1]);
  srv.request.b = atoll(argv[2]);
  // instantiate the service class and assign values to its request
  members

  if (client.call(srv)) // call service
  {
    ROS_INFO("Sum: %ld", (long int)srv.response.sum);
  }
  else
  {
    ROS_ERROR("Failed to call service add_two_ints");
    return 1;
  }
```

```
    return 0;
}
```
Finally, compile the node. Modify the CMakeLists.txt file by adding the
following statement to the end of the file.
```
add_executable(example_server src/example_server.cpp)
target_link_libraries(example_server ${catkin_LIBRARIES})
add_dependencies(example_server ch3_pkg_gencpp)

add_executable(example_client src/example_client.cpp)
target_link_libraries(example_client ${catkin_LIBRARIES})
add_dependencies(example_client ch3_pkg_gencpp)
```
Back to workspace, and compile it.
```
$ cd ~/robook_ws
$ catkin_make
```

3.2.13   Writing a Simple Server and Client (Python Implementation)

In this section, we write simple Server and Client nodes in Python.

First, write the Server node, which does the same thing as the previous
section, which will use AddTwoInts.srv.
Go to the previously created ch3_pkg/scripts directory and create the
example_server.py file.
```
$ cd ~/robook_ws/src/ch3_pkg/scripts
$ touch example_server.py
$ gedit example_server.py
```
Use the following code.
```
#! /usr/bin/env python

from ch3_pkg.srv import *
import rospy

def handle_add_two_ints(req):
  print "Returning [%s + %s = %s]"%(req.a, req.b, (req.a + req.b))
  return AddTwoIntsResponse(req.a + req.b)

def add_two_ints_server():
  rospy.init_node('add_two_ints_server') # Initialize the node,
declare the service

  s = rospy.Service('add_two_ints', AddTwoInts, handle_add_two_ints)
  '''
  Declare a service named add_two_ints, using the AddTwoInts service
type.
  All requests are passed to the handle_add_two_ints function for
processing,
  and pass the instance of AddTwoIntsRequest and return the instance of
```

```
AddTwoIntsResponse
  ' ' '

  print "Ready to add two ints."
  rospy.spin() # Keep the code from exiting until the service is closed

if __name__ == "__main__":
  add_two_ints_server()
```
Modify permissions to be executable.
```
$ chmod +x example_server.py
```
Next, write the client node. In the ch3_pkg/scripts directory, create the example_client.py file.
```
$ touch example_client.py
$ gedit example_client.py
```
Use the following code.
```
#! /usr/bin/env python

import sys
import rospy
from ch3_pkg.srv import *

def add_two_ints_client(x, y):
  rospy.wait_for_service('add_two_ints')
  try:
    add_two_ints = rospy.ServiceProxy('add_two_ints', AddTwoInts)
    # Create a handle for calling the service

    resp1 = add_two_ints(x, y)
    return resp1.sum
  except rospy.ServiceException, e:
    print "Service call failed: %s"%e

def usage():
  return "%s [x y]"%sys.argv[0]

if __name__ == "__main__":
  if len(sys.argv) == 3:
    x = int(sys.argv[1])
    y = int(sys.argv[2])
  else:
    print usage()
    sys.exit(1)
  print "Requesting %s+%s"%(x, y)
  print "%s + %s = %s"%(x, y, add_two_ints_client(x, y))
```
Modify permissions to be executable.
```
$ chmod +x example_client.py
```

3.2.14 Testing a Simple Server and Client

Next, we test the written Server and Client.

```
First, run the following command.
$ roscore
Next, run Server.
$ rosrun ch3_pkg example_server (C++)
$ rosrun ch3_pkg example_server.py (Python)
The output can be seen as follows.
Ready to add two ints.
Run Client with the parameters to be calculated.
$ rosrun ch3_pkg example_client 2 8 (C++)
$ rosrun ch3_pkg example_client.py 2 8 (Python)
As can be seen, the Server terminal is as follows.
# C++
[ INFO] [1548628167.111254451]: request: x=2, y=8
[ INFO] [1548628167.111280342]: sending back response: [10]

# python
Returning [2 + 8 = 10]
The Client terminal is as follows.
# C++
[ INFO] [1548628167.111396602]: Sum: 10

# python
Requesting 2+8
2 + 8 = 10
```

In this chapter, we learn the basic usage of ROS, including how to write your own project package, how to write a simple publisher and subscriber, and how to write a simple Server and Client. The contents of this chapter are very important, and all future functional implementations are closely related to the contents of this chapter, so it is recommended that the reader be proficient in it.

Exercises

1. Change the example of the movement of the small turtle in this chapter so that the turtle moves in a straight line along a square trajectory.
2. Modify the message publisher and subscriber examples to change the published message to the current time.
3. Modify the Server and Client examples to create a service that returns the current time.
4. Briefly describe the difference between a message and a service.

Further Reading

1. Wikipedia. ROS/catkin/conceptual_vrview [EB/OL]. http://wiki.ros.org/catkin/conceptual_
   overview.
2. Wikipedia. ROS/catkin/workspace [EB/OL] http://wiki.ros.org/catkin/workspaces.
3. Wikipedia. ROS/catkin/Tutorials [EB/OL] http://wiki.ros.org/catkin/Tutorials.
4. YoonSeok Pyo, HanCheol Cho, RyuWoon Jung, et al. ROS robot programming [M]. ROBOTIS
   Ltd, 2017.
5. Hu, Chunxu. ROS robot development practice [M]. Beijing: machinery industry publishing
   house, 2018.
6. Wikipedia. ROS/catkin/Tutorials/CreatingPackage [EB/OL]. http://wiki.ros.org/catkin/Tutorials/
   CreatingPackage.

Chapter 4
ROS Debugging

In the previous chapter, we have had a preliminary understanding of the framework and basic usage of ROS and learned how to program the basic functions of ROS, such as basic nodes, messages and services. During the development process, it is inevitable that various software/hardware problems will be encountered and the program needs to be debugged. ROS provides a large number of commands and tools to help developers debug their code. In order to solve various problems encountered during development, these commands can help developers understand information about nodes, topics, etc. This chapter will mainly introduce the commands and tools commonly used for ROS debugging and summarize the basic ROS commands, which are very helpful for solving the problems encountered during debugging robots, and developers should be proficient in these commands.

4.1 Common Commands for ROS Debugging

This section will introduce the common ROS debugging commands according to the type of debugging object to give you an initial understanding of the role and usage of these commands.

1. rosnode

   This command serves to output node-related information and is used as follows.

   $ rosnode ping   Test connectivity to node
   $ rosnode list   List active nodes
   $ rosnode info   Print information about node
   $ rosnode machine   List nodes running on a particular machine or list machines
   $ rosnode kill   Kill a running node
   $ rosnode cleanup   Purge registration information of unreachable nodes

© The Author(s), under exclusive license to Springer Nature Singapore Pte Ltd. 2023
F. Duan et al., *Intelligent Robot*, https://doi.org/10.1007/978-981-19-8253-8_4

2. rqt_graph

For viewing the publish-subscribe relationship between nodes in ROS, the ellipse represents the node and the directed edge represents the publish-subscribe relationship between the nodes at its two ends. Note that rqt_graph is itself a node.

```
$ rosrun rqt_graph rqt_graph
```

3. rostopic

The rostopic command is used to get information about ROS topics.

$ rostopic bw	Display bandwidth used by topic
$ rostopic delay	Display delay for topic which has header
$ rostopic echo	Print messages to screen
$ rostopic find	Find topics by type
$ rostopic hz	Display publishing rate of topic
$ rostopic info	Print information about active topic
$ rostopic list	Print information about active topics
$ rostopic pub	Publish data to topic
$ rostopic type	Print topic type

4. rosservice

This command is used to display information about the ROS service.

$ rosservice args	Print the arguments
$ rosservice call	Call the service with the provided args
$ rosservice find	Find services by service type
$ rosservice info	Print information about service
$ rosservice list	List active services
$ rosservice type	Print service type
$ rosservice uri	Print service ROSRPC uri

5. rosparam

Data on the ROS Parameter Server (Parameter Server) can be stored and manipulated via the rosparam command.

$ rosparam set	Set parameter
$ rosparam get	Get parameter
$ rosparam load	Load parameters from file
$ rosparam dump	Dump parameters to file
$ rosparam delete	Delete parameter
$ rosparam list	List parameter names

6. rosed

rosed is used to edit files in ROS.

```
$ rosed [package_name] [filename]
```

7. rosmsg
    This command is used to output information about ROS Message.

$ rosmsg show   Display the fields in a ROS message type
$ rosmsg list   List all messages
$ rosmsg md5   Display the md5 sum of the message
$ rosmsg package   List all messages in a package
$ rosmsg packages   List all packages with messages

4.2 Common Tools for ROS Debugging

There are a number of tools that we can use to accomplish debugging-related tasks, and this section describes these common tools.

4.2.1 Using rqt_console to Modify the Debug Level at Runtime

ROS provides a number of tools for managing log messages. There are two separate GUIs in ROS Kinetic: rqt_logger_level is used to set a node or the logging level of the logger, rqt_console is used to visualize, filter, and analyze log messages.

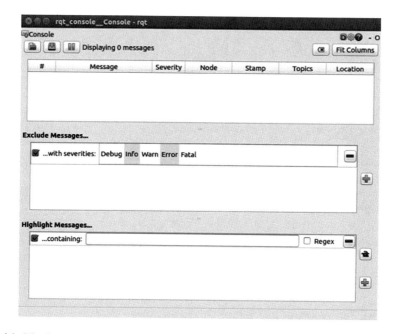

Fig. 4.1 Viewing rqt_console

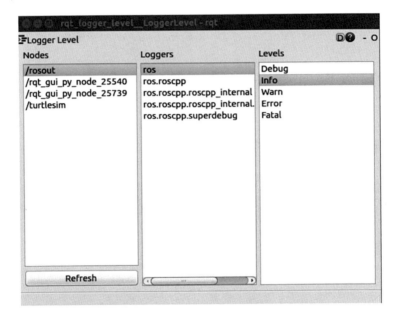

Fig. 4.2 Viewing rqt_logger_level

Let's run roscore and then run the following code under two new terminals to see the output of rqt_console (shown in Fig. 4.1) as well as rqt_logger_level (shown in Fig. 4.2).

```
$ rosrun rqt_console rqt_console
$ rosrun rqt_logger_level rqt_logger_level
```

Now we run turtlesim on the new terminal:

```
$ rosrun turtlesim turtlesim_node
```

If the default log level is Info, you can see the log messages in the window after turtlesim starts, as shown in Fig. 4.3.

Click Refresh to refresh the rqt_logger_level window, and then select Warn to change the logging level to Warn, as shown in Fig. 4.4.

Then, we get the turtle moving by entering the following command on the new terminal.

```
$ rostopic pub /turtle1/cmd_vel geometry_msgs/Twist -r 1 -- '{linear:
{x: 2.0, y: 0.0, z: 0.0}, angular: {x: 0.0,y: 0.0,z: 0.0}}'
```

Observe the output in rqt_console, as shown in Fig. 4.5.

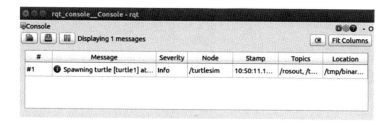

Fig. 4.3 Log messages after turtlesim starts

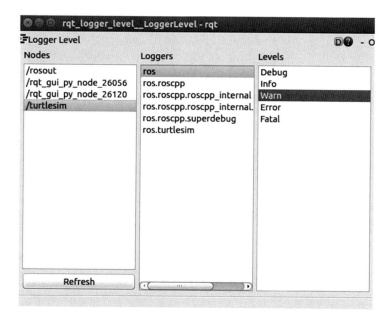

Fig. 4.4 Changing the log level to Warn

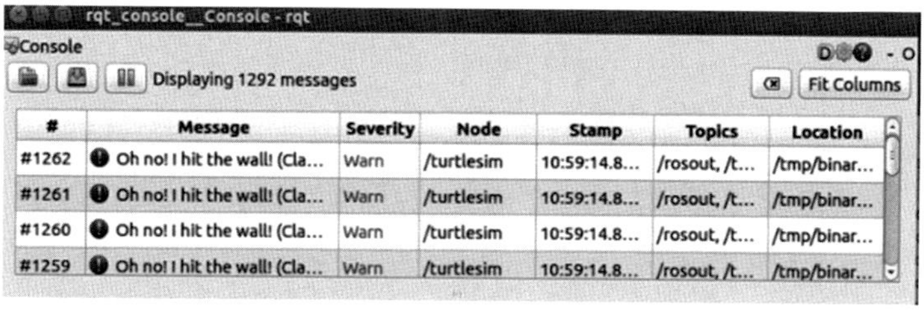

Fig. 4.5 Output in rqt_console

In rqt_console, messages are collected and displayed by category (e.g., by timestamp, message type, severity level, and the node that generated them, etc.). Double-clicking on a message also shows all relevant information, including the code that generated it.

In rqt_logger_level, logs are divided into five priority levels, Fatal, Error, Warn, Info, and Debug, with the highest priority being Fatal and the lowest priority being Debug. By setting the log level, you can get all log messages of priority levels above that level. For example, when setting the log level to Warn, you will get all the log messages of the three levels Warn, Error, and Fatal.

4.2.2 Using roswtf to Detect Potential Problems in the Configuration

ROS provides tools to detect potential problems with all components in a given project package. roswtf is used to check ROS settings (such as environment variables) and find configuration problems. The main uses of roswtf are as follows.

1. For detecting project packages. Use roscd to move to the path of the project package you want to analyze, then run roswtf:

```
$ cd robook_ws
$ source devel/setup.bash
$ roscd ch3_pkg
$ roswtf
```

2. For detecting the launch file, run the following command.

```
$ cd robook_ws
$ source devel/setup.bash
$ roscd ch3_pkg/launch
$ roswtf example_launch.launch
```

The results of the run are shown in Fig. 4.6.

4.2.3 Displaying Node State Graphs Using rqt_graph

In ROS, a directed graph can be used to display the current state of ROS session where the running nodes are the nodes in the graph and the edges are the sender-subscribers, which connect between those nodes and the topics. We run the program that controls turtle with the keyboard, running the following command in each of the three terminals.

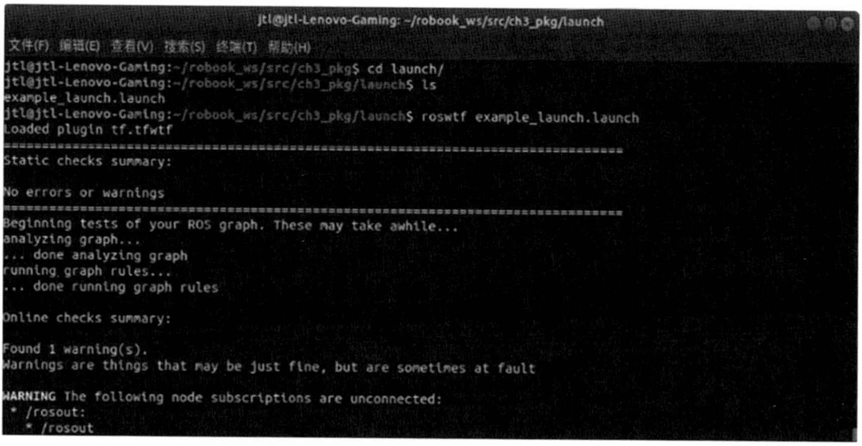

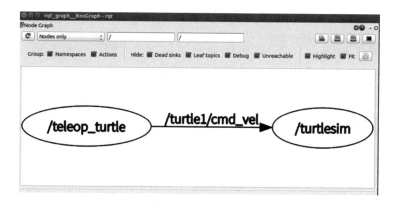

Fig. 4.6 Results of roswtf runs

Fig. 4.7 Node state diagram

```
$ roscore
$ rosrun turtlesim turtlesim_node
$ rosrun turtlesim turtle_teleop_key
```

We then open the status map (shown in Fig. 4.7) at the new terminal.

```
$ rosrun rqt_graph rqt_graph
```

The turtlesim node and the teleop_turtle node communicates with each other through a ROS topic. teleop_turtle nodes publish the input messages on a topic named /turtle1/cmd_vel, and turtlesim nodes receive that message by subscribing to that topic. If you mouse over /turtle1/cmd_vel, the corresponding ROS node and topic are highlighted so that the communication relationship between the publishing and subscribing nodes on this topic is clearly visible.

4.2.4   Plotting Scalar Data Using rqt_plot

In ROS, scalar data can be easily plotted using a number of common tools. rqt_plot provides a GUI plugin that uses a different plotting backend to visualize values in a 2D plot. Here, we will still use the program for keyboard control of the turtle as an example.

From each of the three terminals, run the following command to implement keyboard control of the turtle.

```
$ roscore
$ rosrun turtlesim turtlesim_node
$ rosrun turtlesim turtle_teleop_key
Run rqt_plot and open the GUI interface:
$ rqt_plot
View the running ros topic:
$ rostopic list
```

You can see the following running topics: /rosout, /rosout_agg, /turtle1/cmd_vel /turtle1/color_sensor, /turtle1/pose.

We can see the list of optional topics by typing "/" at the Topic in the top left corner of the opened GUI interface to add (or remove) the topic to be monitored. For example, we monitor the topic /turtle1/pose and then use the keyboard to control the turtle movement to see the change of the parameter curve in the GUI interface (as shown in Fig. 4.8).

4.2.5   Displaying 2D Images Using image_view

In ROS, a node is usually created in which the images from the camera are displayed. If using Microsoft's Kinect camera, we need to install the ROS OpenNI driver. Use the following command.

```
$ sudo apt-get install ros-melodic-openni-camera
```

In this section, we describe how to call a regular USB camera or the one that comes with the laptop. First, we download the ROS driver package for the USB camera from GitHub:

```
$ cd robook_ws/src
$ git clone https://github.com/ros-drivers/usb_cam
```

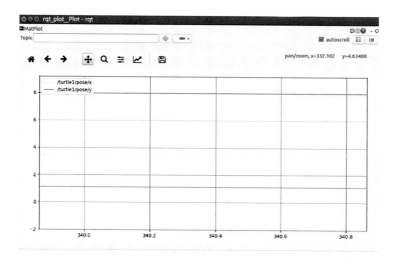

Fig. 4.8 Variation of parameters in the GUI interface

```
$ cd ..
$ catkin_make
```

We get the usb_cam package under robook_ws/src and can view the camera driver with the following command.

```
$ ls /dev/video*
```

We can change the camera device, screen size, etc., by modifying the contents of the usb_cam-test.launch file under the usb_cam package. Now, we run the launch file in the driver package.

```
$ roslaunch usb_cam usb_cam-test.launch
```

There may be some errors, but they don't affect the camera. We run rostopic list in a new terminal, and we get the output of all current topics.

```
$ rostopic list
/rosout
/rosout_agg
/usb_cam/camera_info
/usb_cam/image_raw
/usb_cam/image_raw/compressed
/usb_cam/image_raw/compressed/parameter_descriptions
/usb_cam/image_raw/compressed/parameter_updates
/usb_cam/image_raw/compressedDepth
/usb_cam/image_raw/compressedDepth/parameter_descriptions
/usb_cam/image_raw/compressedDepth/parameter_updates
```

Fig. 4.9   The results of running rqt_image_view

```
/usb_cam/image_raw/theora
/usb_cam/image_raw/theora/parameter_descriptions
/usb_cam/image_raw/theora/parameter_updates
```

We can pick the image information we want to get from it for output. For example, to get the /usb_cam/image_raw topic image information, you can manually select the desired topic image information via rqt_image_view in the top left corner of the open screen. Another example is to select /usb_cam/image_raw/ compressedDepth will give us the depth information. Please note that a normal USB camera does not have depth information. We select /usb_cam/image_raw to get the current color image (as shown in Fig. 4.9).

```
$ rosrun rqt_image_view rqt_image_view
```

4.2.6   3D Data Visualization Using rqt_rviz (rviz)

Many sensor devices (binocular cameras, Kinect, LIDAR, etc.) are capable of providing 3D information. To visualize this data, ROS provides the rviz (rqt_rviz) tool so that sensor data can be displayed in a modeled world. rviz integrates an OpenGL interface capable of 3D data processing, which allows the sensor

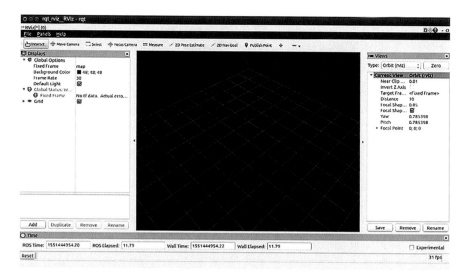

Fig. 4.10 Graphical work interface

coordinate system to read and then plot these readings in the correct position according to the correspondence between the coordinates.

Run roscore and rqt_rviz in two separate terminals.

```
$ roscore
$ rosrun rqt_rviz rqt_rviz
```

We are presented with a graphical work interface shown in Fig. 4.10. The center area of the interface is the 3D view, and on the left is the Display (display type) panel, where all of the loaded project records are displayed. Right now, it contains only the Global Options and Time views. Below this area, there is an Add button for adding more parameter items by topic or type. On the right side of the interface is the View panel, which is used to display the properties of the currently used view controller and the list of saved views.

Each display type has its own status to help the user determine if it is in a normal state. There are generally four statuses: normal, warning, error, and disabled. The status is indicated in the header of the display type by text color, icon, and status category that allows the user to see if the display is expanded. Clicking on each parameter is accompanied by an explanation and description.

There are a number of toolbars at the top of the interface, the most commonly used of which are 2D Pose Estimate and 2D Nav Goal, which can be used in conjunction with the Robot Navigation Engineering package to build the robot's navigation system.

Clicking Add at the bottom brings up the Display Type dialog box, which can be used to add a new display type, as shown in Fig. 4.11. The list at the top of the dialog shows the available display types grouped by the plug-ins that provide them. The

Fig. 4.11 Display type
dialog box

text box in the middle provides a description of the selected display type. In the text
box at the bottom of the dialog box, you can specify a name for the displayed new
instance, which defaults to the name of the type.

The different uses of the rviz visualization tool will be described later. Different
uses require different configurations. To learn the various operations of rviz in detail,
you can refer to http://wiki.ros.org/rviz/Tutorials.

4.2.7 Recording and Playing Back Data Using rosbag and rqt_bag

When we want to record data from the robot for analysis and processing, we need to
use the data saving and playback functions. This section will use rosbag to accomplish this operation.

Here, we are still experimenting with a keyboard-controlled turtle by running the following command in each of the three terminals.

```
$ roscore
$ rosrun turtlesim turtlesim_node
$ rosrun turtlesim turtle_teleop_key
```

Pressing the arrow keys under the terminal that activates turtle_teleop_key to the motion state of the turtle. To publish all topics on the current system, we can open a new terminal and execute the following command.

```
$ rostopic list -v
```

Published topics:

```
* /turtle1/color_sensor [turtlesim/Color] 1 publisher
* /turtle1/command_velocity [turtlesim/Velocity] 1 publisher
* /rosout [roslib/Log] 2 publishers
* /rosout_agg [roslib/Log] 1 publisher
* /turtle1/pose [turtlesim/Pose] 1 publisher
```

Subscribed topics:

```
* /turtle1/command_velocity [turtlesim/Velocity] 1 subscriber
* /rosout [roslib/Log] 1 subscriber
```

The published topic messages listed above are the only topic messages that can be recorded and saved to a file, as only published messages can be recorded. teleop_turtle node publishes a message on the /turtle1/command_velocity topic, while the turtlesim node subscribes to that message as input. turtlesim node simultaneously publishes messages on /turtle1/color_sensor and /turtle1/pose topics.

Now you can start recording messages. First, create a temporary directory for recording messages; then, in that directory, run the rosbag record command with the -a option, which serves to record all currently published topic data and save it to a bag file. This is done by opening a new terminal window and executing the following command in the terminal.

```
$ mkdir ~/bagfiles
$ cd ~/bagfiles
$ rosbag record -a
```

Then, go back to the terminal window where the turtle_teleop node is located and use the keyboard to control the turtle to move randomly for about 10 s. Switch to the window where the rosbag record command is running and press Ctrl-C to exit the command. The contents of the ~/bagfiles directory have a file named with the year,

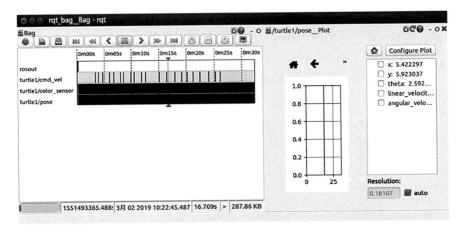

Fig. 4.12 Selecting the topic you want to view

date, and time and with the .bag suffix. This bag file contains the topics posted by all nodes during the run of the rosbag record command.

By entering rosbag info to examine the message content and using the rosbag play command to play it back, we can playback the recorded messages.

```
$ rosbag info <your bagfile>
$ rosbag play <your bagfile>
```

On playback, we are able to see the turtle start moving in the turtuelsim virtual screen like it was previously controlled via the keyboard.

We can also use the rqt_bag command to playback message logging packages, view images, plot scalar data bodies and the RAW structure of messages, and so on. Run the following command and right-click in the blank space of the interface to select the topic you want to view (as shown in Fig. 4.12).

```
$ rosrun rqt_bag rqt_bag <your bagfile>
```

4.2.8   rqt Plugins and rx Applications

Since the release of ROS Fuerte, rx applications or tools have been gradually replaced by rqt nodes, but their functionality is essentially the same, with only a few upgrades and fixes, and some tools are described in Table 4.1.

4.3   Summary of Basic ROS Commands

This section will summarize the basic ROS commands with examples.

Table 4.1 Comparison of ROS rqt and rx tools

ROS rqt tools	ROS rx tools
rqt_console	rxconsole
rqt_graph	rxgraph
rqt_plot	rxplot
rqt_image_view	image_view
rqt_rviz	rviz
rqt_bag	rxbag

4.3.1   Creating a ROS Workspace

Starting up ROS

```
$ roscore
```

Creating and building a catkin workspace

```
$ mkdir -p ~/robook_ws/src
```

```
$ cd ~/robook_ws/src
```

Compiling ROS programs

```
$ cd ~/robook_ws
```

```
$ catkin_make
```

Adding a project package path to environment variable

```
$ echo "source robook_ws/devel/setup.bash" >> ~/.bashrc
```

```
$ source ~/.bashrc
```

4.3.2   Package Related Operations

Create and compile the package

```
$ cd ~/ robook _ws/src
```

```
$ catkin_create_pkg <package_name> [depend1] [depend2] [depend3]
```

```
$ cd ~/ robook _ws
```

```
$ catkin_make
```

Find Package

```
$ rospack find [package name]
```

Search catkin

```
$ catkin_find [package name]
```

View Package dependencies

```
$ rospack depends <package_name>
```

Source-based installation
Find the package to be downloaded from the .rosinstall file, compile and install it,
and generate the setup.bash configuration file to change the environment
variables.

```
$ rosinstall <path> <paths... > [options]
```

Where path is the installation directory of the source code and paths is the .rosinstall
file or its storage directory, which can have more than one.

4.3.3   Related Operations of Nodes

Displays information about ROS nodes

```
$ rosnode list
```

Print information about node

```
$ rosnode info [node_name]
```

Run nodes

```
$ rosrun [package_name] [node_name] [__name=new_name]
```

Test connectivity to specified node

```
$ rosnode ping [node_name]
```

List nodes running on PC

```
$ rosnode machine [PC name or IP]
```

Kill the specified node

```
$ rosnode kill [node_name]
```

Purge registration information of unreachable nodes

```
$ rosnode cleanup
```

4.3.4   Related Operations of the Topic

See all operations of rostopic

```
$ rostopic -h
```

View the list of all topics

```
$ rostopic list
```

Graphical Display Topic

```
$ rosrun rqt_graph rqt_graph
```

```
$ rosrun rqt_plot rqt_plot
```

Display message content of the specified topic in real time

```
$ rostopic echo [topic]
```

Display the message type of the specified topic

```
$ rostopic type [topic]
```

Publish a message with the specified topic name

```
$ rostopic pub [-1] <topic> <msg_type> [--r 1] -- [args] [args]
```

Show topics that use the specified type of message

```
$ rostopic find [topic type]
```

Display the message bandwidth of the specified topic

```
$ rostopic bw [topic]
```

Display the message data release period of the specified topic

```
$ rostopic hz [topic]
```

Display information about the specified topic

```
$ rostopic info [topic]
```

4.3.5   Related Operations of the Service

Display all service operations

```
$ rosservice -h
```

Display a list of current topics

```
$ rosservice list
```

Call the service with the input parameters

```
$ rosservice call [service] [args]
```

Display the service type

```
$ rosservice type [service]
```

Display information about the specified service

```
$ rosservice info [service]
```

Find all services of a specified service type

```
$ rosservice find [service type]
```

Show ROSRPC URI service

```
$ rosservice uri [service]
```

Display the arguments to a service

```
$ rosservice args [service]
```

4.3.6 Related Operations of rosparam

List all parameter names

```
$ rosparam list
```

Save the parameters to the specified file

```
$ rosparam dump [file_name]
```

Set a parameter to a value

```
$ rosparam set [parame_name] [args]
```

Get a parameter value

```
$ rosparam get [parame_name]
```

Load parameters from a specified file

```
$ rosparam load [file_name] [namespace]
```

Delete a parameter value

```
$ rosparam delete [parame_name]
```

4.3.7 Bag-Related Operations

Record all topic

```
$ rosbag record -a
```

Record the specified topic to bag

```
$ rosbag record -O subset <topic1> <topic2>
```

Display the contents of the bag file

```
$ rosbag info <bagfile_name>
```

Play back the contents of specified bag file

```
$ rosbag play (-r 2) <bagfile_name>
```

Compress the specified bag file

```
$ rosbag compress [bagfile_name]
```

Decompress the specified bag file

```
$ rosbag decompress [bagfile_name]
```

Generate a new bag file with the expression removed

```
$ rosbag filter [in-bag] [out-bag] [expression]
```

Reindex the given bag files

```
$ rosbag reindex bag [bagfile_name]
```

Check whether or not the specified bag file is playable in the current system

```
$ rosbag check bag [bagfile_name]
```

Repair the bag file, fix the bag file that can't be played back because of different versions

```
$ rosbag fix [in-bag] [out-bag] [rules]
```

4.3.8   Related Operations of rosmsg

List all messages

```
$ rosmsg list
```

Display the specified message

```
$ rosmsg show [message name]
```

Show md5 sum of the message

```
$ rosmsg md5 [message name]
```

Show all messages for the specified package

```
$ rosmsg package [package name]
```

Show all packages with the message

```
$ rosmsg packages
```

4.3.9   Related Operations of rossrv

List all services

```
$ rossrv list
```

Display information about the specified service

```
$ rossrv show [service name]
```

Show md5 sum of the service

```
$ rossrv md5 [service name]
```

Show all services used in the specified package

```
$ rossrv package [package name]
```

Show all packages with the service

```
$ rossrv packages
```

4.3.10   Other Commands of ROS

Report the disk usage of ROS log files

```
$ rosclean check
```

Remove the corresponding log file

```
$ rosclean purge
```

View ROS Version

```
$ rosversion -d
```

View the version information of the package

```
$ rosversion [package/stack]
```

Check the ROS system

```
$ roswtf
```

Exercise

Check which packages are installed in your ROS directory and list the nodes, messages, and services contained in each package with the command.

Part II
Implementation of Core Robot Functions

Through the first part, we have gained an initial understanding of the use of the ROS robotics operating system and are able to write simple ROS programs and develop and debug them. In this part, we will use some more mature robotics platforms to test common core robot functions, such as robot vision, voice interaction, robotic arm grasping, and other functions.

With the rapid development of information technology such as Internet technology and artificial intelligence technology, the development of robots has developed toward open-source and modularization. In this book, we have modularized the functions of the robot and divided it into visual function module, autonomous navigation function module, voice interaction function module, robot arm control module, etc. For implementing the functions of the various modules of the robot, it is necessary to have a robot platform first. In this book, we use the Turtlebot2 open-source robotics platform. Compared with other robot platforms, Turtlebot2 is easy to operate, highly scalable, and has good compatibility with ROS, which is especially suitable for building ROS test platforms.

In this section, we first introduce the installation and use of the Turtlebot2 robot platform. This is followed by a detailed description of the implementation of each functional module of the robot. Except for the autonomous navigation function, which requires the Turtlebot2 robotics platform, the Turtlebot2 robotics platform is not necessary to implement other functions, but the corresponding functional hardware is required. For example, a Kinect or Primesens camera is required for the vision function; for the voice function, a microphone and audio are required in addition to a laptop (you can use the microphone and audio that come with the laptop); for the control of the robot arm, a servo controller and a robot arm entity are required, etc.

In Chap. 5, we cover the use of the Turtlebot2 robot in detail, including hardware composition and configuration. Then, we learn how to install and test the Turtlebot2 software, including startup, keyboard manual control, script control, and Kobuki battery status monitoring. Finally, the extensions to the Turtlebot robot are introduced, i.e., the hardware architecture and software framework on which the various functions of the robot in this book.

Chapter 6 will describe how to use depth vision sensors to implement robot vision. First, we will learn about the functions and features of the commonly used vision sensors Kinect and Primesense. Then, we will install and test the vision sensor drivers and try to run two Kinects or Kinect and Primesense in ROS at the same time. We will also learn to process RGB images in ROS using OpenCV. Finally, we'll learn about the Point Cloud Library (PCL) and how to use it.

Chapter 7 will give a further introduction to the visual capabilities of robots, which we will delve into and implement more advanced applications. Such as, the robot can be made to walk after its owner, recognize its owner's waving and summoning actions, recognize and locate objects, implement face and gender recognition, and even handwritten digit recognition using the TensorFlow library. Some of these functions are implemented using OpenCV programming, and some require PCL.

Chapter 8 will introduce the autonomous navigation function of the robot. Autonomous navigation of a robot involves key technologies such as robot localization and map building, and path planning. For Turtlebot, it is important first to perform a kinematic analysis of the Kobuki base model first. Then understanding the ROS Navigation Packages and learning to configure and use the navigation package on Turtlebot are necessary.

Chapter 9 will introduce the theory underlying robotic speech interaction functions. The underlying theory of robotic speech interaction function includes techniques such as automatic speech recognition, semantic understanding, and speech synthesis. Among them, the speech recognition section introduces acoustic models and recognition methods such as Hidden Markov Models, Gaussian Mixture Models, Deep Neural Networks. What's more, we introduce language models and methods such as N-gram, NNLM, Word2Vec, etc.; the semantic understanding section introduces Seq2seq methods in detail.

Chapter 10 will describe the implementation of the robot's voice interaction functionality. The routines in this book focus on implementing voice interaction using the PocketSphinx speech recognition system. This chapter first introduces the basic hardware needed for speech recognition and then provides an introduction to the PocketSphinx speech recognition system; after that, it details how to install and test PocketSphinx under the Indigo version and how to publish the results of speech recognition through the ROS topic to control the robot to perform the appropriate tasks. If using ROS Melodic, PocketSphinx will be installed differently, as explained in Sect. 10.4.

Chapter 11 describes the implementation of the robot's robotic arm grasping function, mainly including how to implement a robotic arm using USB2Dynamixel to control the Turtlebot-Arm. This chapter will guide the reader to build a simple robotic arm, as well as install and test the dynamixel_motor package, and implement the robotic arm grasping function in ROS, starting with basic operations such as robotic arm hardware assembly, kinematic analysis, and servo ID setting.

Chapter 5
Installation and Initial Use of the Robots

We have gained a preliminary understanding of the basic framework of the ROS robot operating system, and the best way to further understand the mechanics of how the ROS system works is to test it on a physical robot. Currently, most robot platforms on the market are relatively expensive and have uneven scalability, which is not conducive to use by groups such as robotics enthusiasts and students. Turtlebot is an inexpensive, scalable, and extremely ROS-friendly robot platform, and Turtlebot2 was chosen as the test platform for this book. Turtlebot2 is an official ROS-built robotics platform that runs on the Kobuki base, which is well supported by all versions of ROS, and the routines in this book run primarily on the Turtlebot2 robotics entity.

The main topics in this chapter include: the hardware composition and configuration of the Turtlebot2 robot, software installation and testing; how to start the Turtlebot2 robot, how to control the Turtlebot2 robot movement through keyboard manual control and scripting, and how to monitor the Kobuki base battery status; and finally, the extensions of the Turtlebot2 robot , i.e., the hardware architecture and software framework on which the implementation of the various functions of the robot in this book depends. This chapter gives the reader an initial understanding of the Turtlebot2 robot and the ability to manipulate it in a simple way; at the same time, it builds an overall understanding of the hardware structures and software frameworks on which the implementation of the various robot functions covered in subsequent chapters of this book depends.

5.1 Introduction to the Turtlebot Robot

Turtlebot is an inexpensive and open-source robotics development kit, shown in Fig. 5.1. Turtlebot2 is a new generation of open-source robotics development kit on wheels that can be used to build your own robotics projects. The ROS system provides good support for the Turtlebot2 hardware, which can be used to easily

© The Author(s), under exclusive license to Springer Nature Singapore Pte Ltd. 2023
F. Duan et al., *Intelligent Robot*, https://doi.org/10.1007/978-981-19-8253-8_5

Fig. 5.1 Turtlebot2 and its
charging pile

Table 5.1 List of hardware for Turtlebot

Particulars	Explain
Kobuki base	2200 mAh lithium battery
	19 V adapter
	USB cable
	power line
Depth camera	Microsoft Xbox Kinect/Asus Xion Pro Live
Mechanical component	Fixings
	Structural part
Optional accessory	4400 mAh lithium battery
	Automatic charging pile
	Laptop communicating with Turtlebot

implement 2D map navigation, following, and other functions. By default, the
Turtlebot robot mentioned in this book is the Turtlebot2 robot.

5.2   Composition and Configuration of the Turtlebot Robot Hardware

The Turtlebot robot's hardware is shown in Table 5.1. Its hardware structure is
shown in Fig. 5.2.

The Turtlebot comes with two cables: a USB A-B cable for connecting to the
computer and a USB splitter-type cable for connecting to the Kinect. Plug the A side
of the USB cable into the laptop and the B side into the Kobuki dock. Next, plug the
female end of the USB splitter into the cable leading from the Kinect, the male USB
end of the splitter into the laptop, and the other half into the 12 V 1.5 A plug on the
Kobuki dock, as shown in Fig. 5.3.

Fig. 5.2 Turtlebot's hardware components

Fig. 5.3 Kobuki base USB

Fig. 5.4 Kobuki base on/off button

The Power button is a switch on the left (shown in Fig. 5.4) that chirps and the Status LED lights up when turned on.

Generally, laptops use their own power supply, not the Turtlebot's. If you want your laptop to be able to charge instantly on the Turtlebot, please refer to http://learn. turtlebot.com/2015/02/01/18/.

5.3   Installation and Testing of the Turtlebot Robot Software

5.3.1   Installing from Source

This section focuses on the process of installing the Turtlebot package from source. This installation is characterized by the fact that the source code in the package is installed in the user directory (~/), and the files for the run nodes are in ~/rocon, ~/ kobuki, and ~/turtlebot in the corresponding folders, so if you want to change parameters or a node during operation, you can directly find the corresponding file in these folders and make the changes. This is the recommended installation method for this reason.

1. Preparation before installation

   Before installation, the following command should be execute.

   ```
   $ sudo apt-get install python-rosdep python-wstool ros-indigo-ros
   $ sudo rosdep init
   $ rosdep update
   ```

2. install rocon, kobuki, turtlebot

These three are in relation to the combination space and must be installed in order.

```
$ mkdir ~/rocon
$ cd ~/rocon
$ wstool init -j5 src https://raw.github.com/robotics-in-concert/
rocon/kinetic/rocon.rosinstall
$ source /opt/ros/indigo/setup.bash
$ rosdep install --from-paths src -i -y
$ catkin_make

$ mkdir ~/kobuki
$ cd ~/kobuki
$ wstool init src -j5 https://raw.github.com/yujinrobot/yujin_tools/
master/rosinstalls/indigo/kobuki.rosinstall
$ source ~/rocon/devel/setup.bash
$ rosdep install --from-paths src -i -y
$ catkin_make

$ mkdir ~/turtlebot
$ cd ~/turtlebot
$ wstool init src -j5 https://raw.github.com/yujinrobot/yujin_tools/
master/rosinstalls/indigo/turtlebot.rosinstall
$ source ~/kobuki/devel/setup.bash
$ rosdep install --from-paths src -i -y
$ catkin_make
```

The hierarchy of the three workspaces can also be seen in the order of installation above and in the order of source's setup.bash file.

For convenience, the setup.sh script can be obtained from .bashrc so that the environment is ready when the user logs in.

```
# For a source installation
$ echo "source ~/turtlebot/devel/setup.bash" >> ~/.bashrc
```

5.3.2 Deb Installation Method

This section will describe the deb install method. This installation is simple enough to execute the apt-get command, but installing it this way results in many packages in the turtlebot workspace existing separately in /opt/ros/indigo/share, and it is impossible to see how these packages relate to each other. If you want to run a node individually but don't remember the name of the node, you also can't find the node based on the code, because there are countless packages in the /opt/ros/indigo/ share folder and we simply can't find the node.

Therefore, installing the turtlebot package as a deb is not recommended for beginners who want to understand the relationship between packages in ROS's workspace, which nodes the launch file launches, and how to write their own nodes and launch files.

In fact, the deb install is the apt-get install, with the following command.

```
$ sudo apt-get install ros-indigo-turtlebot ros-indigo-turtlebot-apps
ros-indigo-turtlebot-interactions ros-indigo-turtlebot-simulator
ros -indigo-kobuki-ftdi ros-indigo-rocon-remocon ros-indigo-rocon-
qt-library ros-indigo-ar-track-alvar-msgs
```

If you just want to use some of the packages in turtlebot and don't care about the code, then just execute this command above.

For convenience, the setup.sh script can be obtained from .bashrc so that the environment is ready after login:.

```
# For a source installation
> echo "source ~/turtlebot/devel/setup.bash" >> ~/.bashrc
```

5.3.3 Configuration According to Kobuki Base

1. Set Kobuki alias
   The Kobuki dock used by Turtlebot2 needs to transmit a udev rule to the system so that the system can detect the built-in ftdi USB chip so that the system can read the /dev/kobuki device instead of the unreliable

   ```
   /dev/ttyUSBx device.
   ```

   If it is installed as source, execute the following command.

   ```
   # From the devel space
   > . ~/turtlebot/devel/setup.bash
   > rosrun kobuki_ftdi create_udev_rules
   ```

   If it is installed as a deb, execute the following command.

   ```
   > . /opt/ros/indigo/setup.bash
   > rosrun kobuki_ftdi create_udev_rules
   ```

2. Optional 3D sensor configuration
   By default, Turtlebot Indigo software is used with Asus Xtion Pro. If there are Kinect, Realsense or Orbbec Astra cameras, but they are not configured to be used by Turtlebot, then they need to be set up as described below. This part of the setup

is divided into two scenarios and needs to be chosen by yourself depending on the installation method.

If it is installed as source, the following command should be execute.

```
>echo "export TURTLEBOT_3D_SENSOR=<sensor_name>" >> ~/turtlebot/
devel/setup.sh
> source ~/turtlebot/devel/setup.bash
```

Here, <sensor_name> should be replaced with the name of the corresponding camera, such as kinect, r200 or astra.

If installed as a deb, the following command should be execute.

```
> export TURTLEBOT_3D_SENSOR=<sensor_name>
```

Again, the <sensor_name> above should be replaced with the corresponding camera name, such as kinect, r200, or astra. for convenience, this directive can also be added to the system ~/.bashrc file or to setup.sh in the workspace.

Before executing these commands you need to ensure that the system has the correct robot description details and drivers loaded (Indigo, Kinect and Asus require different drivers) and that the appropriate ROS boot file is called at runtime.

For Kinect, you first need to install OpenNI SDK and Kinect sensor module on Ubuntu; for Asus Xtion Pro, you need to install OpenNI2 SDK, please refer to Sect. 6.2 for the installation and configuration method.

5.4   Launching Turtlebot

Before starting the Turtlebot, you should connect your laptop to the Turtlebot with a USB cable. Press the power button of the dock. Open a new terminal in Ubuntu and enter the following command.

```
$ roslaunch turtlebot_bringup minimal.launch --screen
```

After a successful start, the Turtlebot will chirp and the Status LED will light up in green.

If the startup fails, an error message may appear. Common error prompts and the reasons for them are described below.

```
Already Launched
minimal.launch may have been started.
Kobuki does not start
If the following warning is received.
[ WARN] [1426225321.364907925]: Kobuki : device does not (yet)
available, is the usb connected?
```

```
[ WARN ] [1426225321.615097034] : Kobuki : no data stream, is kobuki
turned on?
```

Obviously, you need to make sure that Kobuki is turned on (the LED on Kobuki is lit to indicate it is on) and that the cable is plugged into the laptop. If these two tasks have been completed, then check to see if the system has applied the udev rule to /dev/kobuki. The check command is as follows.

```
$ ls -n /dev | grep kobuki
```

If it does not exist, run the following script to copy the entire udev rule and restart udev (requires the sudo password).

```
rosrun kobuki_ftdi create_udev_rules
Trying to start Kobuki, not Create
```

If you are using the Create platform, make sure to pre-set the appropriate environment variables.

```
Serial Port - Permission Denied
```

Add the user to the dialout group. this usually only affects users on the Create platform. the /dev/ttyUSBx port requires the user to be in the dialout group. If the user is not in this group, Turtlebot will fail with permission denied. The command to add your user to this group.

```
sudo adduser username dialout
Netbook suspended or powered off when lid was closed
```

Go to dash in the upper left corner of your laptop and type power, then click the power settings icon. When running on AC power and battery, set it so that when the laptop lid is closed it does not perform any
   Operation.

5.5   Manual Control of Turtlebot via Keyboard

The keyboard and the joystick are the two ways to achieve manual operation. Only the use of the keyboard to control the Turtlebot is covered in this book. To learn how to operate the joystick remotely, refer to http://learn.turtlebot.com/2015/02/01/9/, or http://wiki.ros.org/turtlebot_teleop/Tutorials/indigo/Joystick% 20Teleop.
   First, we need to start Turtlebot to receive the command.

```
$ roslaunch turtlebot_bringup minimal.launch
```

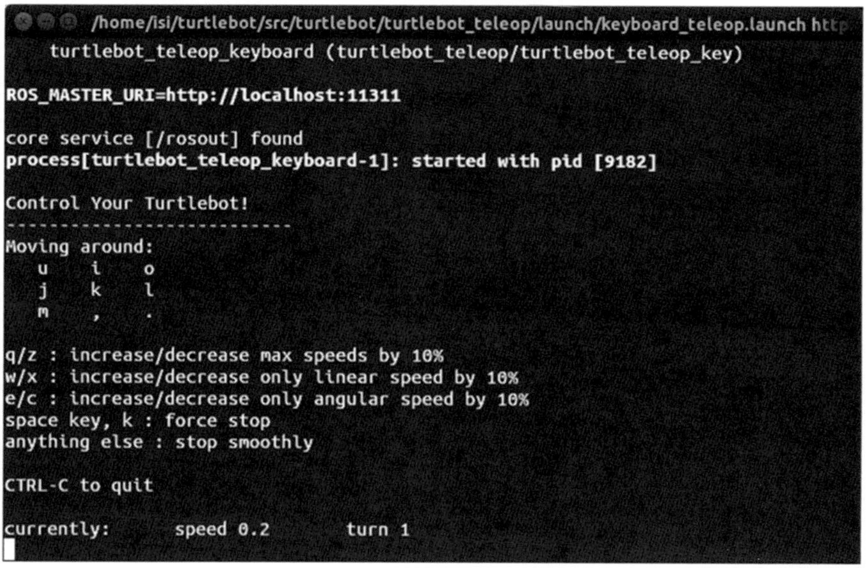

Fig. 5.5   Controlling Turtlebot from the keyboard

To use the keyboard for this, the following command needs to be execute in a new terminal.

```
$ roslaunch turtlebot_teleop keyboard_teleop.launch
```

The terminal displays keyboard commands for manual control of the robot's position, rotation, etc., as shown in Fig. 5.5.

5.6   Controlling Turtlebot Through Scripting

First, find the /turtlebot_hello folder in the book's resource kit and copy it to ~/, or download the sample script from GitHub with the following command.

```
$ mkdir ~/turtlebot_hello
$ cd ~/turtlebot_hello
$ git clone https://github.com/markwsilliman/turtlebot/
```

To control Turtlebot with the Python script downloaded above, you need to start Turtlebot first:.

```
$ roslaunch turtlebot_bringup minimal.launch
```
1. To make the robot go straight, enter the following command in a new terminal.

```
$ cd ~/turtlebot_hello/turtlebot
$ python goforward.py
```

Now the Turtlebot can move forward, press Ctrl+c to stop the forward movement.

2. To make the robot turn in a circle, enter the following command in a new terminal.

```
$ cd ~/ turtlebot_hello/turtlebot
$ python goincircles.py
```

Now Turtlebot is ready to spin around, press Ctrl+c to stop the spin.

We can view/edit the script with the gedit command and can change the velocity and angular velocity by modifying the linear.x and angular.z variables.

Note: Turtlebot only uses linear.x and angular.z because it works in a flat (2D) world, but for drones, marine robots and other robots in 3D environments, Linear.y, Angular.x and Angular.y can be used.

3. If you want the robot's travel path to be a square, enter the following command in a new terminal.

```
$ cd ~/turtlebot_hello/turtlebot
$ python draw_a_square.py
```

Now, the Turtlebot starts moving along a square path on the floor, but it gets off the starting point due to sliding, imperfect calibration, and other factors that cause the robot to not move precisely enough.

5.7 Monitoring the Battery Status of the Kobuki

First, start Turtlebot with the following command.

```
$ roslaunch turtlebot_bringup minimal.launch
```

To monitor the battery status of the Kobuki, enter the following command in a new terminal.

```
$ cd ~/turtlebot_hello/turtlebot
$ python kobuki_battery.py
```

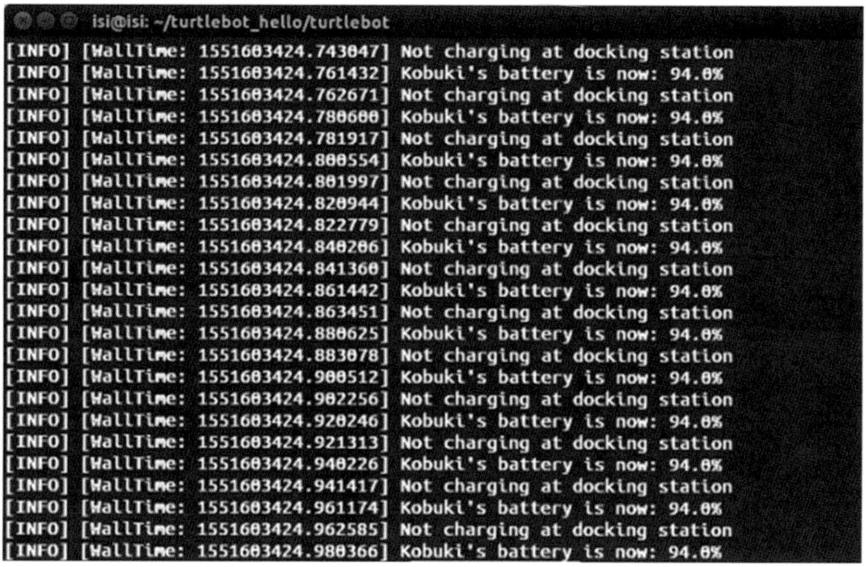

```
isi@isi: ~/turtlebot_hello/turtlebot
[INFO] [WallTime: 1551603424.743047] Not charging at docking station
[INFO] [WallTime: 1551603424.761432] Kobuki's battery is now: 94.0%
[INFO] [WallTime: 1551603424.762671] Not charging at docking station
[INFO] [WallTime: 1551603424.780600] Kobuki's battery is now: 94.0%
[INFO] [WallTime: 1551603424.781917] Not charging at docking station
[INFO] [WallTime: 1551603424.800554] Kobuki's battery is now: 94.0%
[INFO] [WallTime: 1551603424.801997] Not charging at docking station
[INFO] [WallTime: 1551603424.820944] Kobuki's battery is now: 94.0%
[INFO] [WallTime: 1551603424.822779] Not charging at docking station
[INFO] [WallTime: 1551603424.840206] Kobuki's battery is now: 94.0%
[INFO] [WallTime: 1551603424.841360] Not charging at docking station
[INFO] [WallTime: 1551603424.861442] Kobuki's battery is now: 94.0%
[INFO] [WallTime: 1551603424.863451] Not charging at docking station
[INFO] [WallTime: 1551603424.880625] Kobuki's battery is now: 94.0%
[INFO] [WallTime: 1551603424.883078] Not charging at docking station
[INFO] [WallTime: 1551603424.900512] Kobuki's battery is now: 94.0%
[INFO] [WallTime: 1551603424.902256] Not charging at docking station
[INFO] [WallTime: 1551603424.920246] Kobuki's battery is now: 94.0%
[INFO] [WallTime: 1551603424.921313] Not charging at docking station
[INFO] [WallTime: 1551603424.940226] Kobuki's battery is now: 94.0%
[INFO] [WallTime: 1551603424.941417] Not charging at docking station
[INFO] [WallTime: 1551603424.961174] Kobuki's battery is now: 94.0%
[INFO] [WallTime: 1551603424.962585] Not charging at docking station
[INFO] [WallTime: 1551603424.980366] Kobuki's battery is now: 94.0%
```

Fig. 5.6 Monitoring Kobuki battery status

The Kobuki base uses multiple batteries and the power percentage is calculated in kobuki_battery.py using a fixed kobuki_base_max_charge value. To determine the maximum charge of a battery, fill the Turtlebot battery, leave it for a while, and then execute the following command.

$ rostopic echo /mobile_base/sensors/core

Record the value of the battery, then go to the turtlebot folder under turtlebot_hello and edit kobuki_battery.py with the following command.

$ gedit kobuki_battery.py

Modifying the kobuki_base_max_charge value will now calculate the power percentage, as shown in Fig. 5.6.

There are other programs in the turtlebot_hello folder that interested readers can try one by one by referring to http://learn.turtlebot.com/.

5.8   Extensions to the Turtlebot Robot

The hardware structure and software framework on which the implementation of the various functions of the robots in this book depends is based on the Turtlebot robot design implementation. The case robots in this book can implement autonomous navigation, voice recognition, following, image recognition (face recognition, hand waving recognition, gender recognition), visual servo, and other functions, and are compact, lightweight, open source, intelligent, and user-friendly. Figure 5.7 shows the hardware structure diagram of the intelligent home service robot that this book will lead you to design and develop.

The hardware structure on which the implementation of each function of the intelligent service robot relies consists of the following components.

1. Mobile base: Use Turtlebot's Kobuki two-wheel differential base, which has two driven wheels at the front and rear for stable robot movement.
2. Voice system: Receive sound signal through microphone, make sound through audio, and use voice recognition technology to obtain the owner's intention and send it to other functional programs to perform tasks.
3. Dual RGBD cameras: Considering that the objects and faces to be recognized may be located at a high position, and the common obstacles in the home environment are usually located at a low position, the intelligent home service robot designed in this book uses two RGBD cameras at the same time, with the upper part realizing the functions of following, face recognition, waving recognition, and object recognition through the Primesense RGBD camera, and the bottom part realizing the functions of autonomous navigation through the Kinect

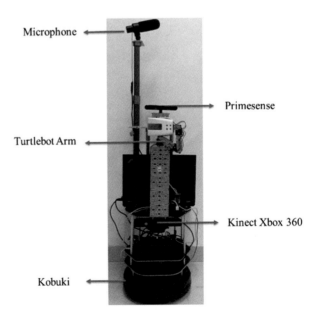

Fig. 5.7 Hardware structure of an intelligent home service robot developed based on ROS

Fig. 5.8 Robot software
architecture design. (**a**)
Encapsulation of robot
hardware. (**b**) Packages
communicate through nodes
and topics

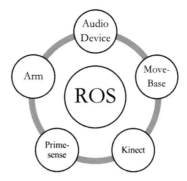

a) Encapsulation of robot hardware

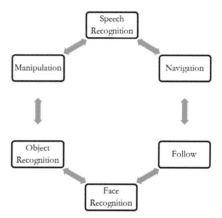

b) Packages communicate through nodes and topics

RGBD camera to achieve autonomous navigation functions. Due to the distributed architecture using ROS, the two cameras can each do their own job and do not affect each other, increasing the spatial range of image recognition, following and autonomous navigation to avoid obstacles better, and providing a powerful vision system for the robot.

4. Robotic arm: The Turtlebot-Arm robotic arm on the robot's chest consists of five Dynamixel AX-12A servos and several 3D printed structural components. The robotic arm can perform grasping operations by inverse kinematic operations based on object coordinates, or perform some actions as needed.

5. Computer: we used a ThinkPad E460 laptop with Ubuntu 14.04 LTS installed, using the ROS version Indigo.

ROS encapsulates the hardware of the robot to implement different hardware control or data flow processing, and interacts with information through a kind of ROS-based peer-to-peer communication mechanism to make the modules of the robot work more efficiently and flexibly, as shown in Fig. 5.8a. In the robot design of this book, the hardware system mainly consists of Kobuki mobile base, Kinect vision sensor, Primesense vision sensor, Turtlebot-Arm robot arm, and voice capture device, and the signals of hardware control, data flow, and processing results are mainly interacted with topics through nodes. To facilitate sharing and distribution, the processes are grouped according to functional packages (Package) or sets of functional packages (Stack). In this book, we mainly design the navigation function package set, face recognition function package, object recognition function package, depth information based following function package, robotic arm grasping control function package, and speech recognition function package. Each functional package of the robot carries out information transfer with the topic through the runtime process (node), thus realizing the collaborative operation of each functional part of the robot, as shown in Fig. 5.8b.

This chapter describes the software, hardware, and basic usage of the Turtlebot robotics platform. This chapter gives us a general idea of the robotics platform we will be using next, which helps to understand the implementation of the different functions in later chapters and how to use the various sensors, robotic arms, and other hardware.

Exercise
Run the programs in the turtlebot_hello folder in sequence to see how they run.

Further Reading

1. IMYUNE. ros robot [EB/OL]. http://turtlebot.imyune.com/.
2. Learn TurtleBot and ROS. Hardware Setup [EB/OL]. http://learn.turtlebot.com/2015/02/01/3/.
3. Wikipedia.ROS:PCInstallation [EB/OL]. http://wiki.ros.org/turtlebot/Tutorials/indigo/PC%20 Installation.
4. Wikipedia. ROS:turtlebot Bringup [EB/OL]. http://wiki.ros.org/turtlebot_bringup/Tutorials/ indigo/Turtlebot%20Bringup.
5. Learn TurtleBot and ROS. Writing Your First Script [EB/OL]. http://learn.turtlebot.com/201 5/02/01/10/.

Chapter 6
Implementation of Robot Vision Functions

Robot vision relies on the corresponding vision sensors for implementation, and the commonly used sensors are cameras with image capabilities. In order to achieve stereo vision, cameras with depth information perception are also required. The robot vision system in this book is designed with dual RGBD vision sensors. The upper part uses Primesense RGBD camera for following, face recognition, hand waving recognition and object recognition, and the lower part uses Kinect RGBD camera for autonomous navigation function.

This chapter focuses on how to use Kinect and Primesense cameras for vision functions in ROS system. First, we introduce the features and uses of the vision sensors Kinect and Primesense; then we learn how to install and test the drivers of these two sensors; then we try how to run two Kinects in ROS at the same time, and how to run Kinect and Primesense at the same time, how to use OpenCV in ROS to process RGB images; and finally, we introduce the Point Cloud Library (PCL) and how to use it.

Through the study and practice of this chapter, the reader should have a preliminary understanding of Kinect and Primesense and be able to run both sensors simultaneously. In addition, the reader should be able to use OpenCV to process RGB images in ROS and be able to use PCL for point cloud data processing. This knowledge is the basis for subsequent implementation of autonomous robot navigation, following, face and object recognition and needs to be mastered.

6.1 Vision Sensors

The main functions of vision sensors include image acquisition, image processing, image recognition (face recognition, hand waving recognition, object recognition) and vision servo. Vision sensors for service robots also need to have the function of obtaining depth information. Commonly used RGBD cameras include Kinect, Primesense, Astra, etc., whose working principles are basically the same. They are

© The Author(s), under exclusive license to Springer Nature Singapore Pte Ltd. 2023
F. Duan et al., *Intelligent Robot*, https://doi.org/10.1007/978-981-19-8253-8_6

inexpensive, and are ideal for developing robot vision functions. In this section, two commonly used vision sensors-Kinect vision sensor and Primesense vision sensor-are introduced.

6.1.1 Kinect Vision Sensor

Microsoft's Kinect is a commonly used visual sensor. The Kinect Xbox 360, for example, is equipped with an RGB camera, a depth sensor, and a microphone array with a lens in the center. The depth sensor consists of an infrared laser projector on the left and a monochrome CMOS depth sensor on the right that captures 3D video data under any lighting condition, as shown in Fig. 6.1. The infrared emitter emits infrared light, which is reflected to the infrared receiver after touching an object. kinect Xbox 360 uses Light Coding (Light Coding) measurement technology. When an object reflects infrared light, the structure of the light changes with the distance of the object, and the infrared receiver resolves and calculates the distance of the object based on the structure of different light.

The Kinect's various sensors output video at a frame rate of 9–30 Hz depending on the resolution. The default RGB video stream uses 8-bit VGA resolution (640 × 480 pixels) and a Bayesian color filter, but the resolutions supported by hardware up to 1280 × 1024 pixels (at lower frame rates) and other color formats such as UYVY. The monochrome CMOS depth sensor video stream uses VGA resolution (640 × 480 pixels) with 11-bit depth and provides 2048 level of sensitivity. The Kinect depth sensor has a practical application range of 1.2–3.5 m. The Kinect sensor has an angular field of view of 57° horizontally and 43° vertically, and the middle rotation axis can tilt the sensor up or down to 27°.

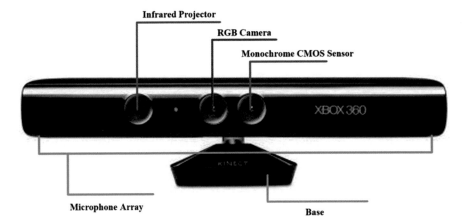

Fig. 6.1 Kinect Xbox 360 camera

6.1.2 Primesense Vision Sensors

Another vision sensor often used in vision recognition systems for service robots is the Primesense 1.09 RGBD vision sensor, shown in Fig. 6.2, which is mainly used for following, waving recognition, face recognition, object recognition, and other functions. The Primesense 1.09 is similar to the Kinect in that it can provide an RGB camera, depth sensor, and microphone array, and supports USB power, but it is smaller and lighter, making it easier to integrate on top of the robot. The sensor has an infrared light source on the left side, a depth CMOS image sensor in the middle, and an RGB camera on the right side. Primesense 1.09 offers 16-bit VGA (640 × 480 pixels) RGB video streaming and depth video streaming with Light Coding (Light Coding) measurement technology. It has a small range of 0.35–1.4 m, which is suitable for close view recognition.

Table 6.1 gives a comparison of the attribute parameters of Primesense 1.09 and Kinect Xbox 360. As can be seen, Primesense 1.09 has the following features compared to Kinect Xbox 360.

1. Primesense 1.09 is easy to integrate on top of the robot due to its compact size and light weight, reducing the burden of the robot's base motion.
2. The use of Primesense 1.09 short-range camera allows object detection and recognition in the neighborhood, thus circumventing the influence of the complex

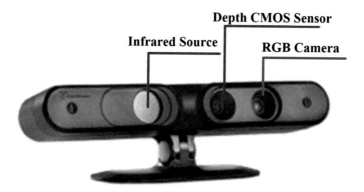

Fig. 6.2 Pimesense 1.09 RGBD vision sensor

Table 6.1 Comparison of attribute parameters for Primesense 1.09 and Kinect Xbox 360

Genus	Primesense 1.09	Kinect Xbox 360
Volume (excluding base)	18 cm × 2.5 cm × 3.5 cm	28 cm × 4 cm × 6 cm
Vision System	Depth + RGB Camera	Depth + RGB Camera
Ranging technology	Optical encoding technology	Optical encoding technology
Audio System	Microphone Array	Microphone Array
Power/Interface	USB2.0/3.0	External Power + USB2.0
Distance measurement	0.35–1.4 m	1.2–3.5 m

background of the home environment on object recognition and improving the recognition accuracy.

3. Primesense 1.09 needs to provide the spatial coordinates of the object to be grasped for the robot arm, and performs visual servo for the "grasp" and "release" of the robot arm. Since the robot arm in this book is relatively short, the only way to meet the requirements is to use a short-depth camera to match the working space of the robot arm.

In summary, Primesense close up camera is generally suitable for short distance recognition scenes within 1m, such as following, face recognition, object recognition and other scenes; Kinect camera is suitable for slightly farther scenes in the range of 2–5 m, such as navigation and other scenes.

6.2   Driver Installation and Testing

1. Install openni and freenect drivers
   The command to install the openni and freenect drivers is as follows.

   ```
   $ sudo apt-get install ros-indigo-openni-* ros-indigo-openni2-*
   ros-indigo-freenect-*
   $ rospack profile
   ```

2. Set environment variables
   First, check the environment variables and determine the output of Turtlebot's default 3D sensor with the following command.

   ```
   $ echo $TURTLEBOT_3D_SENSOR
   #Output: kinect
   ```

   If it is another 3D sensor, e.g. output is asus_xtion_pro, you need to set the default value of the environment variable. Run the following command to change TURTLEBOT_3D_SENSOR to kinect and restart the terminal.

   ```
   $ echo "export TURTLEBOT_3D_SENSOR=kinect" >> .bashrc
   ```

3. Start the camera
   Execute the following command from the Turtlebot terminal.

   ```
   $ roslaunch turtlebot_bringup minimal.launch
   ```

   In Turtlebot, open a new terminal and run different commands depending on the version.

1) For Microsoft Kinect, run the following command.

```
$ roslaunch freenect_launch freenect-registered-xyzrgb.
launch
```

If it is an older version of Microsoft Kinect, run the following command.

```
$ roslaunch freenect_launch freenect.launch
$ roslaunch openni_launch openni.launch
```

2) For the Asus Xtion/Xtion Pro/Primesense 1.08/1.09 cameras, run the following command.

```
$ roslaunch openni2_launch openni2.launch
depth_registration:=true
```

4. Test camera

To test whether the camera can display images, you can open a terminal and execute the following command.

```
$ rosrun image_view image_view image:=/camera/rgb/image_raw
```

6.3   Running Two Kinects at the Same Time

1. Operating environment

Before implementing two Kinects running simultaneously, you need to prepare the following running environments: Kinect Xbox, Ubuntu 14.04, ROS Indigo, ThinkPad (with more than 2 USB BUS, not PORT).

2. Implementation steps

First, run the following command.

```
$ lsusb
```

The terminal window will output the following result.

```
Bus 002 Device 001: ID 1d6b:0003 Linux Foundation 3.0 root hub
Bus 001 Device 004: ID 04f2:b541 Chicony Electronics Co., Ltd
Bus 001 Device 003: ID 8087:0a2a Intel Corp.
Bus 001 Device 010: ID 045e:02ae Microsoft Corp. Xbox NUI Camera
Bus 001 Device 008: ID 045e:02b0 Microsoft Corp. Xbox NUI Motor
Bus 001 Device 009: ID 045e:02ad Microsoft Corp. Xbox NUI Audio
```

```
Bus 001 Device 007: ID 0409:005a NEC Corp. HighSpeed Hub
Bus 001 Device 002: ID 17ef:6050 Lenovo
Bus 001 Device 014: ID 045e:02ae Microsoft Corp. Xbox NUI Camera
Bus 001 Device 012: ID 045e:02b0 Microsoft Corp. Xbox NUI Motor
Bus 001 Device 013: ID 045e:02ad Microsoft Corp. Xbox NUI Audio
Bus 001 Device 011: ID 0409:005a NEC Corp. HighSpeed Hub
Bus 001 Device 001: ID 1d6b:0002 Linux Foundation 2.0 root hub
```

From the output, it appears that there are two Xbox devices. Run the following command.

```
$locate freenect.launch
Get the location of the file freenect.launch, usually in /opt/ros/
indigo/share/freenect_launch/launch/
In freenect.launch.
$ cd /opt/ros/indigo/share/freenect_launch/launch/
```

Then, we need to write our own launch file: the

```
$ sudo gedit doublekinect_test.launch
```

The document reads as follows.

```
<launch>
<!-- Parameters possible to change -->
<arg name="camera1_id" default="#1" /><!--here you can change 1@0
by the serial number -->
<arg name="camera2_id" default="#2" /><!--here you can change 2@0
by the number -->
<!--arg name="camera1_id" default="B00366600710131B" /--><!--
here you can change 1@0 by the serial number -->
<!--arg name="camera2_id" default="B00364210621048B" /--><!--
here you can change 2@0 by the serial number -->
<!--arg name="camera3_id" default="#3" /--><!--here you can
change 3@0 by the serial number -->
<arg name="depth_registration" default="true"/>

<!-- Default parameters -->
<arg name="camera1_name" default="kinect1" />
<arg name="camera2_name" default="kinect2" />
<!--arg name="camera3_name" default="kinect3" /-->

<!-- Putting the time back to realtime -->
<rosparam>
/use_sim_time : false
</rosparam>

<!-- Launching first kinect -->
<include file="$(find freenect_launch)/launch/freenect.launch">
<arg name="device_id" value="$(arg camera1_id)"/>
```

```
  <arg name="camera" value="$(arg camera1_name)"/>
  <arg name="depth_registration" value="$(arg depth_registration)" /
>
  <node name="rviz" pkg="rviz" type="rviz"/>
  </include>
  <! -- Launching second kinect -->
  <include file="$(find freenect_launch)/launch/freenect.launch">
  <arg name="device_id" value="$(arg camera2_id)"/>
  <arg name="camera" value="$(arg camera2_name)"/>
  <arg name="depth_registration" value="$(arg depth_registration)" /
>
  <node name="rviz" pkg="rviz" type="rviz"/>
  </include>
  <! -- Launching third kinect -->
  <! --include file="$(find openni_launch)/launch/openni.launch" -->
  <! --arg name="device_id" value="$(arg camera3_id)"/-->
  <! --arg name="camera" value="$(arg camera3_name)"/-->
  <! --arg name="depth_registration" value="$(arg
depth_registration)" /-->
  <! --/include -->
  </launch>
```

Save the results.

Note: between <! -- --> are comment messages, which can be ignored.

If the computer is configured with more than one USB BUS, you can connect
3 or even more Kinects.

3. Testing

First, open a terminal and run the following command.

```
$roslaunch freenect_launchdoublekinect_test.launch
```

Open another terminal and run the following command.

```
$rosrun image_view image_view image:=/kinect1/rgb/image_color
```

This will bring up the depth map of the first Kinect. Open another terminal and
run the following command.

```
$ rosrun image_view image_view image:=/kinect2/rgb/image_color
```

This will bring up a depth map of the second Kinect, as shown in Fig. 6.3. As
you can see, we were successful in running both Kinects at the same time.

Fig. 6.3 Running two kinects at the same time

6.4   Running Kinect and Primesense at the Same Time

To run Kinect and Primesense at the same time, the following environment needs to be prepared.

Kinect Xbox, Ubuntu 14.04, ROS Indigo, ThinkPad (with 2+ USB BUS, not PORT).

Next, run the following command.

```
$ sudo -s
```

Otherwise the following warning will appear.

```
Warning: USB events thread - failed to set priority. This might cause
loss of data...
```

Then run Primesense with the following command.

```
$ roslaunch openni2_launch openni2.launch camera=camera1
```

Or modify the camera parameter in the openni2.launch file and save it as a new file, primesense_test.launch. i.e., change the

```
<arg name="camera" default="camera"/>
```

Amend to read.

```
<arg name="camera" default="camera1"/>
$ roslaunch openni2_launch primesense_test.launch
```

Open a new terminal and run Kinect:.

```
$ roslaunch freenect_launch freenect.launch
```

At this point, you can see that both Kinect and Primesense are running successfully.

Note: Primesense has to be started here first.

6.5 RGB Image Processing With OpenCV in ROS

6.5.1 Installing OpenCV in ROS

OpenCV (Open Source Computer Vision Library) is a cross-platform computer open source vision library that implements many common algorithms in the field of image processing and computer vision. OpenCV provides a large number of image processing functions, including image display, pixel manipulation, target detection, etc., which greatly simplifies the development process.

ROS generally comes with the latest stable version of OpenCV, which can be viewed with the command pkg-config --modversion opencv. As shown in Fig. 6.4, the version of OpenCV queried on ROS Indigo is 2.4.8. Post-ROS Kinetic installations generally default to OpenCV3, but OpenCV3 is not packaged into debian/ubuntu.

ROS Indigo comes with OpenCV2, and ROS after Indigo comes with the OpenCV package, which can be installed by searching for OpenCV3 in the Ubuntu Software Center, or by using the following command.

```
$ sudo apt-get install ros-indigo-opencv3
```

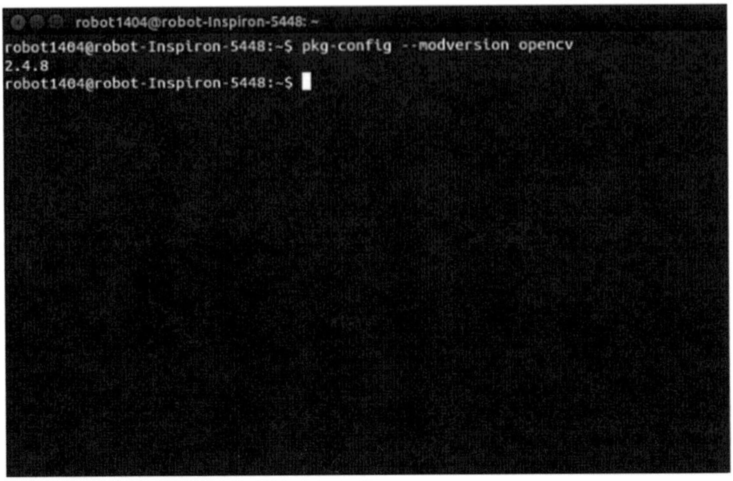

Fig. 6.4 Querying OpenCV versions

6.5.2   Using OpenCV in ROS Code

OpenCV2 is the official version supported by Indigo and Jade. To use it, we just need to add a dependency on OpenCV2 in CMakeLists.txt and set the find_package entry, just like for any other third-party package.

```
find_package(OpenCV)
  include_directories(${OpenCV_INCLUDE_DIRS})
  target_link_libraries(my_awesome_library ${OpenCV_LIBRARIES})
```

We can also use OpenCV3, in which case dependencies need to be added to OpenCV3. But you should make sure that all dependencies are not dependent on OpenCV2 (because the link between both versions of OpenCV at the same time can create conflicts).

If OpenCV2 and ROS OpenCV3 are already installed, find_package will first find OpenCV3.

In this case, if you want to compile with OpenCV2, set find_package as follows.

```
find_package(OpenCV 2 REQUIRED)
```

6.5.3 Understanding the ROS-OpenCV Conversion Architecture

ROS transfers images in sensor_msgs/Image message format, but users prefer to manipulate image data through data types or objects, and OpenCV is the most commonly used library for this. In OpenCV, unlike the image message format defined by ROS ,images are stored as Mat matrices, , so we need to link the two different formats by using CvBridge, a ROS library that provides an interface between ROS and OpenCV.The conversion architecture for ROS and OpenCV is shown in Fig. 6.5.

Next, we detail the conversion of ROS to OpenCV using the C++ language as an example. If you use other languages or platforms, please refer to http://wiki.ros.org/cv_bridge/Tutorials.

1. Convert ROS image to OpenCV image

CvBridge defines a CvImage class containing OpenCV image variables and their encoding variables, and ROS image header variables. CvImage in turn contains information about sensor_msgs/Image and can be converted between them. The CvImage class is defined as follows.

```
namespace cv_bridge {

  class CvImage
  {
    public:
      std_msgs::Header header;//ROS image Header variable
      std::string encoding;//OpenCV image encoding variable
      cv::Mat image;//OpenCV image variable
  };
```

Fig. 6.5 ROS-OpenCV conversion architecture

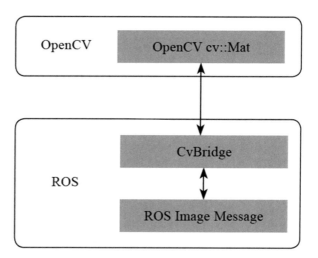

```
typedef boost::shared_ptr<CvImage> CvImagePtr;
typedef boost::shared_ptr<CvImage const> CvImageConstPtr;
```

```
}
```

When converting ROS sensor_msgs / Image messages to CvImage, CvBridge will recognizes two different use cases.

1) We want to modify the data in-place, then we must copy the ROS message data.
2) We don't want to modify the data, then we can safely share the data owned by the ROS message instead of copying it.

CvBridge provides the following functions for conversion to CvImage.

```
// Scenario 1: always copy, return CvImage of variable type
CvImagePtr toCvCopy(const sensor_msgs::ImageConstPtr& source,
        const std::string& encoding = std::string());
CvImagePtr toCvCopy(const sensor_msgs::Image& source,
        const std::string& encoding = std::string());
```

```
// Scenario 2: Share as much as possible, return CvImage of constant
type
    CvImageConstPtr toCvShare(const sensor_msgs::ImageConstPtr&
source,
        const std::string& encoding = std::string());
CvImageConstPtr toCvShare(const sensor_msgs::Image& source,
        const boost::shared_ptr<void const>& tracked_object,
        const std::string& encoding = std::string());
```

The input is a message pointer to the image, and an optional encoding parameter. The encoding is the specified CvImage.

toCvCopy creates a copy of the image data from the ROS message, even if the source and target encodings match. However, we are free to modify the returned CvImage.

If the source and target encodings match, toCvShare will point the returned cv::Mat to the ROS message data to avoid copying. As long as a copy of the returned CvImage is kept, the ROS message data is not freed. If the encoding does not match, it will allocate a new buffer and perform the conversion. Modification of the returned CvImage is not allowed, as it may share data with ROS image messages, which in turn may share data with other callbacks.

Note: It is more convenient to make use of the second overloaded function of toCvShare when there are pointers to other message types (e.g. stereo_msgs/ DisparityImage) that contain the sensor_msgs/Image to be converted.

If no encoding is provided (or more precisely, an empty string), the destination image encoding will be the same as the image message encoding. In this case, toCvShare guarantees that no image data will be copied.

For a detailed explanation of the above functions and the image encoding issues, please refer to http://wiki.ros.org/cv_bridge/Tutorials/ UsingCvBridgeToConvertBetweenROSImagesAndOpenCVImages.

2. OpenCV image conversion to ROS image message

We can use the toImageMsg() member function to convert a CvImage to a ROS image message as follows.

```
class CvImage
{
  sensor_msgs::ImagePtr toImageMsg() const;

  // Overload is mainly used for aggregate messages containing the
  member sensor msgs::image.
  void toImageMsg(sensor_msgs::Image& ros_image) const;
};
```

If the CvImage is assigned by the user, then don't forget to fill in the header and encoding fields. For instructions on how to assign instances of CvImage, please refer to the following tutorials: http://wiki.ros.org/image_transport/Tutorials/ PublishingImages (publish images); http://wiki.ros.org/image_transport/Tuto rials/SubscribingToImages (subscrib to images); http://wiki.ros.org/image_ transport/

Tutorials/ExaminingImagePublisherSubscriber (run the image publisher and receiving server).

6.5.4   ROS Node Example

This section will show an example of a node. The node listens to the ROS image message topic, converts that image to cv::Mat format, and then uses OpenCV to draw a circle on the image and display it. The image will then be republished in ROS.

1. Create a project package for image processing

First, if you have already created the project package, you need to add the following dependencies to package.xml and CMakeLists.xml.

```
sensor_msgs
cv_bridge
roscpp
std_msgs
image_transport
Or to create a project package by using catkin_create_pkg.
$ cd ~/robook_ws/src
```

```
$ catkin_create_pkg ch6_opencv sensor_msgs cv_bridge roscpp
std_msgs image_transport
```

2. Edit node

Create the image_converter.cpp file in the ch6_opencv/src folder and add the following.

```
 1 #include <ros/ros.h>
 2 #include <image_transport/image_transport.h>
 3 #include <cv_bridge/cv_bridge.h>
 4 #include <sensor_msgs/image_encodings.h>
 5 #include <opencv2/imgproc/imgproc.hpp>
 6 #include <opencv2/highgui/highgui.hpp>
 7
 8 static const std::string OPENCV_WINDOW = "Image window";
 9
10 class ImageConverter
11 {
12   ros::NodeHandle nh_;
13   image_transport::ImageTransport it_;
14   image_transport::Subscriber image_sub_;
15   image_transport::Publisher image_pub_;
16
17 public:
18   ImageConverter() : it_(nh_)
19
20   {
21     // Subscrive to input video feed and publish output video feed
22     image_sub_ = it_.subscribe("/camera/rgb/image_color", 1,
23        &ImageConverter::imageCb, this);
24     image_pub_ = it_.adaptise("/image_converter/output_video",
       1);
25
26     cv::namedWindow(OPENCV_WINDOW);
27   }
28
29   ~ImageConverter()
30   {
31     cv::destroyWindow(OPENCV_WINDOW);
32   }
33
34   void imageCb(const sensor_msgs::ImageConstPtr& msg)
35   {
36     cv_bridge::CvImagePtr cv_ptr;
37     try
38     {
39       cv_ptr = cv_bridge::toCvCopy(msg, sensor_msgs::
         image_encodings::BGR8);
40     }
41     catch (cv_bridge::Exception& e)
42     {
43       ROS_ERROR("cv_bridge exception: %s", e.what());
44       return;
```

```
45    }
46
47    // Draw an example circle on the video stream
48    if (cv_ptr->image.rows > 60 && cv_ptr->image.cols > 60)
49      cv::circle(cv_ptr->image, cv::Point(50, 50), 10, CV_RGB
        (255,0,0));
50
51    // Update GUI Window
52    cv::imshow(OPENCV_WINDOW, cv_ptr->image);
53    cv::waitKey(3);
54
55    // Output modified video stream
56    image_pub_.publish(cv_ptr->toImageMsg());
57  }
58 };
59
60 int main(int argc, char** argv)
61 {
62   ros::init(argc, argv, "image_converter");
63   ImageConverter ic;
64   ros::spin();
65   return 0;
66 }
```

Let's analyze the code content of the above node.

Line 2: Use image_transport to publish and subscribe to images in ROS, which can subscribe to compressed image streams. The image_transport should be included in package.xml.

Lines 3–4: Contain the header file for CVBridge as well as some useful constants and functions related to image encoding. The cv_bridge should be included in package.xml.

Lines 5–6: Include header files for the OpenCV image processing and GUI modules. The opencv2 should be included in package.xml.

Lines 12–24: Use image_transport to subscribe to image topics as "input" and publish image topics as "output".

Lines 26–32: OpenCV HighGUI requires a call to create/destroy the display window on startup/shutdown.

Lines 34–45: In our subscriber callback, the ROS image message is first converted to a CVImage suitable for use with OpenCV. because we want to draw on the image, we need a mutable copy of it, and toCvCopy() is used.

Note: OpenCV requires that color images toned to use BGR channel rules.
Note: CvCopy() / toCvShared() should be called to catch conversion errors, as those functions do not check the validity of the data.

Lines 47–52: Draw a red circle on the image and then display it in the display window.

Line 53: Convert the CvImage to a ROS image message and post it on the "Output" topic.

3. Related preparation and setup

Since it is a C++ program, the following needs to be added to CMakeLists.txt.

```
add_executable(image_converter src/image_converter.cpp) //add
the file in src as an executable file with the name image_converter
    target_link_libraries(image_converter ${catkin_LIBRARIES}) //
link the relevant libraries and executable files
    add_dependencies(image_converter
robot_vision_generate_messages_cpp) //add dependency packages to
the executable file
```

To run this node, an image stream needs to be generated. We can run the camera or play the package file to generate the image stream. For example, using the Kinect camera:

```
$ roslaunch openni_launch openni.launch
```

4. Compile and run

We run the following command to compile:

```
$ cd ~/robook_ws
$ catkin_make
$ source devel/setup.bash
$ rosrun ch6_opencv image_converter
```

You can also use roslaunch, which eliminates the need to open a separate Kinect. launch file reads:

```
<launch>
    <include file="/opt/ros/indigo/share/openni_launch/launch/
openni.launch" />
    <node name="image_converter" pkg="ch6_opencv"
type="image_converter" output="screen" >
    </node>
</launch>
```

If you have successfully converted the image to OpenCV format, you will see a HighGui window named "Image window" which displays the image and circle.

We can check if the node is publishing images correctly through ROS by using rostopic or by viewing the image with the use of Image_View. For example:

```
$ rostopic info topic name // show the content of the delivered topic
```

perhaps

```
$ rosrun image_view image_view image:=/image_converter/
output_video
```

For the usage of toCvShare() for sharing images, please refer to: http://wiki. r o s . o r g / c v _ b r i d g e / T u t o r i a l s / UsingCvBridgeToConvertBetweenROSImagesAndOpenCVImages.

6.6   Point Cloud Library and Its Use

6.6.1   Introduction to Point Clouds and Point Cloud Libraries

Point Cloud data is a collection of vectors in a 3D coordinate system. These vectors are usually represented in the form of X, Y, Z coordinates and are used to represent the external surface shape of an object. In addition to the geometric position information represented by (X, Y, Z), the point cloud data can also represent the RGB color, depth, gray value, segmentation result, etc. of the points. For example, if $Pi = \{Xi, Yi, Zi\}$ represents a point in space, then Point Cloud = {P1, P2, P3, …, Pn} represents a set of point cloud data.

Most point cloud data is generated by 3D scanning devices such as Stereo Camera, LIDAR (2D/3D), TOF (Time-Of-Flight) cameras, Light Coding cameras, etc. These devices automatically measure information about the points on the surface of an object and then output the point cloud data with some sort of data file.

The Point Cloud Library (PCL) is a standalone C++ library for 3D point cloud processing.The PCL framework contains a large number of advanced algorithms covering feature estimation, filtering, alignment, surface reconstruction, model fitting and segmentation. The main application areas of PCL are robotics, virtual reality, laser telemetry, CAD/CAM, reverse engineering, human-computer interaction, etc.

PCL supports the Primesense 3D vision sensor and Kinect used in this book.

6.6.2   Data Types of PCL

1. Point cloud data structures in ROS
   The main types of data structures currently used to represent point clouds in ROS are as follows.

   sensor_msgs::PointCloud: the first point cloud message used in ROS. It contains X, Y and Z (all floating point data) and multiple channels, each with a string name and an array of floating point values.

sensor_msgs::PointCloud2: The latest revision of the ROS point cloud message (currently the actual standard for PCL). It now represents arbitrary n-dimensional data. Point values can now be of any basic data type (int, float, double, etc.) and messages can be specified as dense, with height and width values, giving the data a two-dimensional structure. For example, we can correspond to an image of the same region in space.

pcl::PointCloud<T>: The core point cloud class in the PCL library, which can be templated on any of the point types listed in point_types.h or user-defined types. This class has a similar structure to the PointCloud2 message type, including the header. Direct conversions between message classes and point cloud template classes are possible (seeing the introduction below), and most methods in the PCL library accept objects of both types. However, it is preferable to use this template class in point cloud processing nodes rather than using message objects, since individual points can be used as objects without having to use their original data.

2. Common PointCloud2 field names

Commonly used Point Cloud2 field names are as follows.

x: the value of the x coordinate of a point (float32).

y: the value of the y coordinate of a point (float32).

z: the value of the z coordinate of a point (float32).

rgb: the RGB (24-bit) color value of a point (uint32).

rgba: the A-RGB (32-bit) color value of a point (uint32).

normal_x: the first component of the normal direction vector of a point (float32).

normal_y: the second component of the normal direction vector of a point (float32).

normal_z: the third component of the normal direction vector of a point (float32).

curvature: an estimate of the change in surface curvature of a point (float32).

j1: the first invariant moment of a point (float32).

j2: the second invariant moment of a point (float32).

j3: the third invariant moment of a point (float32).

boundary_point: the boundary property of the point (e.g., set to 1 if the point lies on a boundary, boolean).

principal_curvature_x: the first component of the principal curvature direction of a point (float32).

principal_curvature_y: the second component of the principal curvature direction of a point (float32).

principal_curvature_z: the third component of the principal curvature direction of a point (float32).

A full list of field names and point types used in PCL can be found in pcl/point_types.hpp. Please refer to http://docs.ros.org/hydro/api/pcl/html/point__types_8hpp.html.

3. Point cloud data type conversion

The conversion between sensor_msgs::PointCloud2 and pcl::PointCloud<T> objects can be done by using pcl::fromROSMsg and pcl::toROSMsg in pcl_conversions.

The easiest way to convert the formats between sensor_msgs::PointCloud and sensor_msgs::PointCloud2 is to run a node instance of point_cloud_converter (http://wiki.ros.org/point_cloud_ converter). This node subscribes to both types of topics and publishes both types of topics. To convert in your own node, please refer to sensor_msgs::convertPointCloud2ToPointCloud and sensor_msgs:: convertPointCloudToPoint-Cloud2.

6.6.3 Publish and Subscrib to Point Cloud Messages

For the sake of completeness, we will summarize the subscription and publishing operations for the three pointcloud types below. Note: We do not promote the use of older pointcloud message types.

1. Subscribe to different point cloud message types
   For all types, the following actions need to be completed.

   ```
   ros::NodeHandle nh;
   std::string topic = nh.resolveName("point_cloud");
   uint32_t queue_size = 1;
   ```

   For the sensor_msgs::PointCloud topic, complete the following.

   ```
   // callback signature:
   void callback(const sensor_msgs::PointCloudConstPtr&);

   // create subscriber:
   ros::Subscriber sub = nh.subscribe(topic, queue_size, callback);
   ```

   For the sensor_msgs::PointCloud2 topic, complete the following.

   ```
   // callback signature
   void callback(const sensor_msgs::PointCloud2ConstPtr&);

   // to create a subscriber, you can do this (as above):
   ros::Subscriber sub = nh.subscribe<sensor_msgs::PointCloud2>
   (topic, queue_size, callback);
   ```

   For subscribers that receive pcl::PointCloud<T> objects directly, complete the following.

   ```
   // Need to include the pcl ros utilities
   #include "pcl_ros/point_cloud.h"
   ```

```
// callback signature, assuming your points are pcl::
PointXYZRGB type:
  void callback(const pcl::PointCloud<pcl::PointXYZRGB>::
ConstPtr&);

// create a templated subscriber
  ros::Subscriber sub = nh.subscribe<pcl::PointCloud<pcl::
PointXYZRGB> > (topic, queue_size, callback);
```

When using the sensor_msgs::PointCloud2 subscriber to subscrib to a pcl::PointCloud<T> topic (and vice versa), the conversion (deserialization) between the two types sensor_msgs::PointCloud2 and pcl::PointCloud<T> will be done immediately by the subscriber.

2. Publish different point cloud types

As with subscriptions, for all types, the following actions need to be completed.

```
ros::NodeHandle nh;
std::string topic = nh.resolveName("point_cloud");
uint32_t queue_size = 1;
```

For sensor_msgs::PointCloud messages, complete the following.

```
// assume you get a point cloud message somewhere
sensor_msgs::PointCloud cloud_msg;

// advertise
ros::Publisher pub = nh.advertise<sensor_msgs::PointCloud>(topic,
queue_size);
// and publish
pub.publish(cloud_msg);
```

For sensor_msgs::PointCloud2 messages, complete the following.

```
// get your point cloud message from somewhere
sensor_msgs::PointCloud2 cloud_msg;

// to advertise you can do it like this (as above):
ros::Publisher pub = nh.advertise<sensor_msgs::PointCloud2>
(topic, queue_size);

/// and publish the message
pub.publish(cloud_msg);
```

For a pcl::PointCloud<T> object, without converting it into a message, the operation is as follows.

```
// Need to include the pcl ros utilities
#include "pcl_ros/point_cloud.h"

// you have an object already, eg with pcl::PointXYZRGB points
pcl::PointCloud<pcl::PointXYZRGB> cloud;

// create a templated publisher
ros::Publisher pub = nh.advertise<pcl::PointCloud<pcl::
PointXYZRGB> > (topic, queue_size);

// and just publish the object directly
pub.publish(cloud);
```

The publisher is responsible for the conversion (serialization) between sensor_msgs::PointCloud2 and pcl::PointCloud<T> when needed.

6.6.4   Tutorial on How to Use PCL in ROS

This section will describe how to use existing tutorials on http://pointclouds.org (using nodes or node sets) in ROS. There are three main routines included here: example.cpp, example_voxelgrid.cpp and example_planarsegmentation.cpp.

1. Create a ROS project package
   Create a ROS project package by using the following command.

```
$ cd robook_ws/src
$ catkin_create_pkg ch6_pcl pcl_conversions pcl_ros roscpp
sensor_msgs
```

   Then add the following to the package.xml file.

```
<build_depend>libpcl-all-dev</build_depend>
<exec_depend>libpcl-all</exec_depend>
```

2. Create the code framework
   Create the code framework example.cpp as follows.

```
#include <ros/ros.h>
// PCL specific includes
#include <sensor_msgs/PointCloud2.h>
#include <pcl_conversions/pcl_conversions.h>
#include <pcl/point_cloud.h>
#include <pcl/point_types.h>

ros::Publisher pub;
```

```
void cloud_cb (const sensor_msgs::PointCloud2ConstPtr& input)
{
  // Create a container for the data.
  sensor_msgs::PointCloud2 output;

  // Do data processing here...
  output = *input;

  // Publish the data.
  pub.publish (output);
}

int main (int argc, char** argv)
{
  // Initialize ROS
  ros::init (argc, argv, "ch6_pcl");
  ros::NodeHandle nh;

  // Create a ROS subscriber for the input point cloud
  ros::Subscriber sub = nh.subscribe ("input", 1, cloud_cb);

  // Create a ROS publisher for the output point cloud
  pub = nh.advertise<sensor_msgs::PointCloud2> ("output", 1);

  // Spin
  ros::spin ();
}
```

Note: the code above only does the work of initializing the ROS and creating subscribers and publishers for PointCloud2 data.

3. Add the source file to CMakeLists.txt

In the newly created project package edit the CMakeLists.txt file and add the following.

```
add_executable(example src/example.cpp)
target_link_libraries(example ${catkin_LIBRARIES})
```

4. Download the source code from the PCL tutorial

PCL has four different ways to represent point cloud data. Beginners may be a bit confused, but we'll try to make it as easy as possible for everyone to learn. The four types are.

```
sensor_msgs::PointCloud-ROS message (deprecated).
sensor_msgs::PointCloud2-ROS message.
pcl::PCLPointCloud2-PCL data structure, mainly for compatibility
with ROS.
pcl::PointCloud<T> - Standard PCL data structure.
```

In the following code example, we will focus on the ROS message (sensor_msgs::PointCloud2) and the standard PCL data structure (pcl::PointCloud<T>). It should be noted, however, that pcl::PCLPoint-Cloud2 is also an important and useful type with the use of subscribing to nodes directly, and it will automatically interconvert with sensor_msgs types via serialization. Please refer to the example_voxelgrid_pcl_types.cpp file in the resources of this book, or refer to the following link: http://wiki.ros.org/pcl/Tutorials/hydro?action=AttachFile&do=view&target=example_voxelgrid_pcl_types.cpp to learn the usage of PCLPointCloud2 by yourself.

(1) sensor_msgs/PointCloud2

The source files for the following examples can be downloaded by using 6.6.4example_voxelgrid.cpp in the book's resources or on: http://wiki.ros.org/pcl/Tutorials/hydro?action=AttachFile&do=view&target=example_voxelgrid.cpp. Note that cmakelists.txt has to be edited to match.

sensor_msgs::/PointCloud2 is designed for ROS messages and is the preferred format for ROS applications. In the following example, we have simplified the PointCloud2 structure by using a 3D grid, which greatly reduces the number of points in the input data set.

To add this functionality to the code framework above, perform the following steps.

1) Get the file 6.6.4voxel_grid.cpp in the book's resources, or go to http://www.pointclouds.org/documentation/, click Tutorials, and navigate to the Downsampling a PointCloud using a VoxelGrid filter tutorial (http://www.pointclouds.org/documentation/tutorials/voxel_grid.php).

2) Read the code and instructions, you can see that the code is divided into three parts: loading the cloud (lines 9–19), processing the cloud (lines 20–24), and saving the output (lines 25–32).

3) Since we are using the ROS subscription server and publishing server in the code snippet above, we can ignore loading and saving the point cloud data with the use of the PCD format. Therefore, the only relevant parts of this tutorial are lines 20–24 where the PCL object is created, input data is passed, and the actual computation is performed, as follows.

```
// Create the filtering object
pcl::VoxelGrid<pcl::PCLPointCloud2> sor;
sor.setInputCloud (cloud);
```

```
sor.setLeafSize (0.01, 0.01, 0.01);
sor.filter (*cloud_filtered);
```

In these lines, the input dataset is named cloud and the output dataset is called cloud_filtered. we can replicate this work, but remember to use the sensor_msgs class instead of the pcl class. In order to do this we need to do some additional work for converting the ROS messages to PCL types. Modify the callback function in the code framework example.cpp as follows.

```
#include <pcl/filters/voxel_grid.h>
```

. . . . . .

```
void
cloud_cb (const sensor_msgs::PointCloud2ConstPtr& cloud_msg)
{
  // Container for original & filtered data
  pcl::PCLPointCloud2* cloud = new pcl::PCLPointCloud2;
  pcl::PCLPointCloud2ConstPtr cloudPtr (cloud);
  pcl::PCLPointCloud2 cloud_filtered;

  // Convert to PCL data type
  pcl_conversions::toPCL(*cloud_msg, *cloud);

  // Perform the actual filtering
  pcl::VoxelGrid<pcl::PCLPointCloud2> sor;
  sor.setInputCloud (cloudPtr);
  sor.setLeafSize (0.1, 0.1, 0.1);
  sor.filter (cloud_filtered);

  // Convert to ROS data type
  sensor_msgs::PointCloud2 output;
  pcl_conversions::fromPCL(cloud_filtered, output);

  // Publish the data
  pub.publish (output);
}
```

Note: Since different tutorials usually use different variable names for input and output, it may be necessary to modify the code when integrating the tutorial code into your own ROS node. In this case, please note that we must change the variable name input to cloud and the output to cloud_filtered to match the code in the tutorial we copied.

(continued)

Note: This code is somewhat inefficient, so we could use moveFromPCL instead of fromPCL to prevent copying the entire (filtered) point cloud. However, since the original input is a constant, it is not possible to optimize the toPCL call in this way.

Save the example.cpp framework as a separate file example_voxelgrid.cpp and make a compilation.

```
$ cd robook_ws
$ catkin_make
```

Next, run the following code.

```
$ rosrun ch6_pcl example_voxelgrid input: = /
narrow_stereo_textured /points2
```

If running an OpenNI-compatible depth sensor, try running it as follows.

```
$ roslaunch openni_launch openni.launch
$ rosrun ch6_pcl example_voxelgrid input:=/camera/depth/points
```

The results can be visualized by running rviz.

```
$ rosrun rviz rviz
```

And add "PointCloud2" to rviz. Select camera_depth_frame for the fixed frame (or whatever frame is appropriate for the sensor), and select the output for the PointCloud2 topic. Here you should see a point cloud downsampled by height. For comparison, you can look at the /camera/depth/points topic and look at the amount it has downsampled.

(2) pcl/PointCloud<T>

As with the previous examples, you can use the attached code \ch6_pcl\src \example_planarsegmentation.cpp from the book's resources, or you can download it on http://wiki.ros.org/pcl/Tutorials/hydro?action=AttachFile&do= view&target=example_planarsegmentation.cpp. Note that the cmakelists.txt has to be edited to match.

The pcl::PointCloud<T> format represents the internal PCL point cloud format. For reasons of modularity and efficiency, this format is templated on point types, and PCL provides a list of SSE-aligned templated common types. In the following example, we want to estimate the plane coefficients of the largest plane in the scene.

To add this functionality to the code framework above, perform the following steps.

1) Get the file 6.6.4planar_segmentation.cpp in the book's resources, or go to http://
   www.pointclouds.org/documentation/, click Tutorials, and navigate to the planar
   model segmentation tutorial (http://www.pointclouds.org/documentation/tuto
   rials/planar_segmentation.php).
2) Read the code and instructions, you can see that the code is divided into three
   parts: creating the cloud and filling it with values (lines 12–30), processing the
   cloud (lines 38–56), and writing down the coefficients (lines 58–68).
3) Since we used the ROS subscriber in the code snippet above, we can ignore the
   first step and process the cloud received in the callback directly. Therefore, the
   only relevant part of this tutorial is lines 38–56, which is the part of the code that
   creates the PCL object, passes the input data, and performs the actual
   computation.

```
pcl::ModelCoefficients coefficients;
pcl::PointIndices inliers;
// Create the segmentation object
pcl::SACSegmentation<pcl::PointXYZ> seg;
// Optional
seg.setOptimizeCoefficients (true);
// Mandatory
seg.setModelType (pcl::SACMODEL_PLANE);
seg.setMethodType (pcl::SAC_RANSAC);
seg.setDistanceThreshold (0.01);

seg.setInputCloud (cloud.makeShared ());
seg.segment (inliers, coefficients);
```

In this part of the code, the input dataset is named cloud and is of type pcl::
PointCloud<pcl::PointXYZ>.

The output is represented by a set of point indices containing the plane and
plane coefficients. cloud.makeShared() creates a boost shared_ptr shared pointer
object for the cloud object (refer to the pcl::PointCloud API documentation:
http://docs.pointclouds.org/1.5.1/classpcl_1_1_point_cloud.html#a33ec29ee932
707f593af9839eb37ea17).

Copy the following line into the code framework example.cpp and modify the
callback function.

```
#include <pcl/sample_consensus/model_types.h>
#include <pcl/sample_consensus/method_types.h>
#include <pcl/segmentation/sac_segmentation.h>


......


void
cloud_cb (const sensor_msgs::PointCloud2ConstPtr& input)
{
  // Convert the sensor_msgs/PointCloud2 data to pcl/PointCloud
```

```
pcl::PointCloud<pcl::PointXYZ> cloud;
pcl::fromROSMsg (*input, cloud);

pcl::ModelCoefficients coefficients;
pcl::PointIndices inliers;
// Create the segmentation object
pcl::SACSegmentation<pcl::PointXYZ> seg;
// Optional
seg.setOptimizeCoefficients (true);
// Mandatory
seg.setModelType (pcl::SACMODEL_PLANE);
seg.setMethodType (pcl::SAC_RANSAC);
seg.setDistanceThreshold (0.01);

seg.setInputCloud (cloud.makeShared ());
seg.segment (inliers, coefficients);

// Publish the model coefficients
pcl_msgs::ModelCoefficients ros_coefficients;
pcl_conversions::fromPCL(coefficients, ros_coefficients);
pub.publish (ros_coefficients);
}
```

> Note: We added two conversion steps: a conversion from sensor_msgs/
> PointCloud2 to pcl/PointCloud<T> and a conversion from pcl::
> ModelCoefficients to pcl_msgs::ModelCoefficients. Also, we changed the
> published variables from output to coefficients.

Also, since we are now publishing planar model coefficients instead of point cloud data, the publisher type must be modified. The original type was:

```
// Create a ROS publisher for the output point cloud
pub = nh.advertise<sensor_msgs::PointCloud2> ("output", 1);
```

The changed type is:

```
// Create a ROS publisher for the output model coefficients
pub = nh.advertise<pcl_msgs::ModelCoefficients> ("output", 1);
```

Save the example.cpp framework as a separate file example_planarsegmentation.cpp and add the source file to CMakeLists.txt, then compile and run the code above with the following command.

```
$ rosrun ch6_pcl example_planarsegmentation input:=/
narrow_stereo_textured/points2
```

If running an OpenNI-compatible depth sensor, try running it as follows.

```
$ rosrun ch6_pcl example_planarsegmentation input:=/camera/
depth/points
```

View the output as follows.

```
$ rostopic echo output
```

6.6.5 A Simple Application of PCL - Detecting the Opening and Closing State of a Door

The following is a simple application of PCL where the scenario is that the robot has navigated to the door, but the door is closed. The robot detects the state of the door. If the door is open, by detecting the change in depth of the point cloud, the robot can detect that the door is open and post the message that the door is open.

The hardware used in this example can be either Kinect or Primesense.

You can find 6.6.5door_detect.cpp and 6.6.5door_detect.launch in the resources of this book, put them in the /src and /launch folders of the project package, respectively, and configure CMakeList.txt for them.

> Note: The default vision device in 6.6.5door_detect.launch is Turtlebot's 3dsensor. You can replace the<include file="/home/isi/turtlebot/src/turtlebot/turtlebot_bringup/launch/3dsensor.launch">
> with <include file="/opt/ros/indigo/share/openni2_launch/launch/openni2.launch">, i.e. use Primesense.

Compile and run as follows.

```
$ cd robook
$ catkin_make# to compile after modifying the cpp program
$ source devel/setup.bash
$ roslaunch ch6_pcl door_detect.launch
```

In this chapter, we learned about using depth vision sensors to implement robot vision capabilities. First, we learned about the two commonly used vision sensors, Kinect and Primesense, then we learned to install and test the vision sensor drivers, and tried to run two Kinects in ROS at the same time, or Kinect and Primesense at the same time. In addition, we introduced the use of OpenCV in ROS to process RGB images, learned about the Point Cloud Library (PCL) and how to use it. This lays the foundation for implementing autonomous navigation, following, face and object recognition for robots in the next chapter.

Exercises

1. Display images acquired by Kinect and Primesense cameras in ROS.
2. Use OpenCV in ROS to read the camera image and display it in rviz.
3. Use the Primesense camera to detect the opening and closing status of the door.

Further Reading

1. Cui B. Research on sound source localization based on Kinect microphone array [D]. Zhenjiang: Jiangsu University, 2015.
2. CSDN. device PrimeSense 1.09_driver installation and use [EB/OL]. https://blog.csdn.net/hanshuning/article/details/56845394.
3. Wikipedia. ROS:vision_opencv [EB/OL]. http://wiki.ros.org/vision_opencv.
4. Wikipedia. ROS:opencv3 [EB/OL]. http://wiki.ros.org/opencv3.
5. Wikipedia. Converting between ROS images and OpenCV images (C++) [EB/OL]. http://wiki.ros.org/cv_bridge/Tutorials/UsingCvBridgeToConvertBetweenROSImagesAndOpenCVImages.
6. Wikipedia. ROS:pcl [EB/OL]. http://wiki.ros.org/pcl.
7. Huang Junjun. Streamlined algorithm for real-time 3D reconstruction study of mobile scenes based on PFH and information fusion [D]. Shanghai: Donghua University, 2014.
8. Wang, Liping. Indoor 3D map construction for mobile robot with fused image and depth information [D]. Xiamen: Xiamen University, 2012.
9. Chen, Jiazhou. Simultaneous recognition and modeling of indoor scene objects [D]. Guangzhou: Guangdong University of Technology, 2016.

Chapter 7
Advanced Implementation of Robot Vision Functions

In the previous chapter, we have gained an initial understanding of the implementation of the robot's vision functions, and in this chapter, we will explore in depth and implement more advanced vision functions of the robot, such as having the robot recognize and follow its owner's movement, recognize its owner's waving call action from multiple people, recognize and locate objects, recognize faces and genders within the field of view, and recognize handwritten numbers. Some of the functions will be implemented using OpenCV programming and some will require PCL for implementation. These functions are the basis for implementing intelligent service robots and should be mastered with emphasis.

First create the project package:

```
$ cd robook_ws/src
$ catkin_create_pkg imgpcl sensor_msgs cv_bridge roscpp std_msgs
image_transport pcl_conversions pcl_ros
```

OpenCV and PCL are then configured as described in Chapter 6 in relation to each other.

7.1 Implementation of the Robot Follow Function

7.1.1 Theoretical Foundations

There are various ways to implement service robot follow, for example, follow function can be accomplished using technologies such as ultrasonic module, image recognition, sound source localization and radar. Here, we use 3D vision sensors to implement the follow function of the service robot.

The core of the algorithm that implements the follow function is that the robot looks for objects in its front detection window and seeks the center of mass of the

© The Author(s), under exclusive license to Springer Nature Singapore Pte Ltd. 2023
F. Duan et al., *Intelligent Robot*, https://doi.org/10.1007/978-981-19-8253-8_7

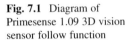

Fig. 7.1 Diagram of
Primesense 1.09 3D vision
sensor follow function

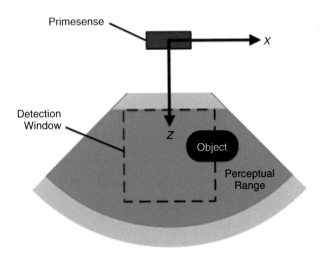

observed object, which preferably remains directly in front of the robot and at a fixed
distance. If the object's center of mass is too far away, the robot will travel forward,
otherwise backward; if the object is offset to the side of the robot, the robot will turn
toward the center of mass. The robot in this book uses the depth camera of
Primesense 1.09 to extract the point cloud information and restrict the point cloud
to a rectangular body with a certain length, width, and height to simulate the amount
of point cloud of the human body, thus excluding the interference of other objects.
The schematic is shown in Fig. 7.1.

The speed control strategy uses proportional control to achieve the effect of
maintaining a certain distance between the robot and the target object and speed
smoothing. The algorithm flow chart is shown in Fig. 7.2.

7.1.2 Operational Testing of Follow Functions

The follow function of the robot can generally be done using two sensors, Kinect or
Primesense. First, the robot needs to be connected to the laptop with the vision
sensor and the robot switch needs to be activated.

You can use Turtlebot's own turtlebot_follower directly to implement the follow
function. First start Turtlebot:

```
$ roslaunch turtlebot_bringup minimal.launch
```

Procedure for starting the follow function at a new terminal.

```
$ roslaunch turtlebot_follower follower.launch
```

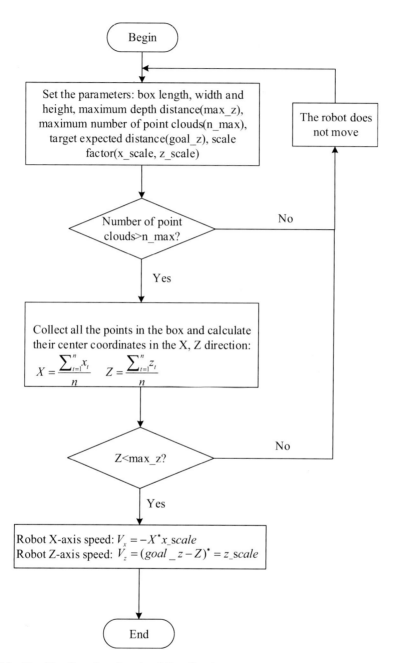

Fig. 7.2 Algorithm flow chart for robot follow function

Note: If there is a problem with the startup of the visual sensor, you need to enter the command sudo - s before running to change the user's permission. The current user's own environment is used here, no user variables need to be loaded, and no directory jumps are required.

The follower.launch file is located under the Turtlebot installation package at turtlebot/src/turtlebot_apps/turtlebot_follower/launch/follower.launch. We can find the vision sensor launch settings in the file at

```
<include file="$(find turtlebot_bringup)/launch/3dsensor.launch">
```

Set which vision sensor to select in this file. When installing Turtlebot, the default vision sensor may have been set.

export TURTLEBOT_3D_SENSOR=kinect

However, for ease of use, the follower.launch can be modified to directly launch the Turtlebot robot and the vision sensors used. The contents of the modified myfollower.launch file are as follows (see the book resource 7.1.2 myfollower.launch):

```
<!--
 The turtlebot people (or whatever) follower nodelet.
 -->
<launch>
<include file="/home/isi/turtlebot/src/turtlebot/turtlebot_bringup/
launch/minimal.launch" />
 <arg name="simulation" default="false"/>
 <group unless="$(arg simulation)"><!-- Real robot -->

  <include file="$(find turtlebot_follower)/launch/includes/
velocity_smoother.launch.xml">
   <arg name = "nodelet_manager" value = "/
mobile_base_nodelet_manager"/>
    <arg name="navigation_topic" value="/cmd_vel_mux/input/navi"/>
   </include>

   <include file="/opt/ros/indigo/share/openni2_launch/launch/
openni2.launch">
    <arg name="rgb_processing" value="true"/>
 <!-- only required if we use android client -->
    <arg name="depth_processing" value="true"/>
    <arg name="depth_registered_processing" value="false"/>
    <arg name="depth_registration" value="false"/>
    <arg name="disparity_processing" value="false"/>
    <arg name="disparity_registered_processing" value="false"/>
   </include>
  </group>
  <group if="$(arg simulation)">
```

```xml
  <!-- Load nodelet manager for compatibility -->
  <node pkg="nodelet" type="nodelet" ns="camera" name =
"camera_nodelet_manager" args="manager"/>

  <include file="$(find turtlebot_follower)/launch/includes/
velocity_smoother.launch.xml">
    <arg name = "nodelet_manager" value = "camera/
camera_nodelet_manager"/>
    <arg name="navigation_topic" value="cmd_vel_mux/input/navi"/>
  </include>
 </group>

 <param name="camera/rgb/image_color/compressed/jpeg_quality" value
= "22"/>

 <!-- Make a slower camera feed available; only required if we use
android client -->
  <node pkg="topic_tools" type="throttle" name="camera_throttle"
    args="messages camera/rgb/image_color/compressed 5"/>

 <include file = "$(find turtlebot_follower)/launch/includes/
safety_controller.launch.xml"/>

 <!-- Real robot: load turtlebot follower into the 3d sensors nodelet
manager to avoid pointcloud serializing -->
 <!-- Simulation: load turtlebot follower into nodelet manager for
compatibility -->
 <node pkg="nodelet" type="nodelet" name="turtlebot_follower"
    args="load turtlebot_follower/TurtlebotFollower camera/
camera_nodelet_manager">
  <remap from = "turtlebot_follower/cmd_vel" to =
"follower_velocity_smoother/raw_cmd_vel"/>
  <remap from="depth/points" to="camera/depth/points"/>
  <param name="enabled" value="true" />
  <param name="x_scale" value="7.0" />
  <param name="z_scale" value="2.0" />
  <param name="min_x" value="-0.35" />
  <param name="max_x" value="0.35" />
  <param name="min_y" value="0.1" />
  <param name="max_y" value="0.5" />
  <param name="max_z" value="1.2" />
  <param name="goal_z" value="0.6" />
 </node>
 <!-- Launch the script which will toggle turtlebot following on and off
based on a joystick button. default: on -->
 <node name="switch" pkg="turtlebot_follower" type="switch.py"/>
 <!-modify:Create new follow.rviz file under turtlebot_follower, load
rviz, at this point rviz content is empty -->
  <node name="rviz" pkg="rviz" type="rviz" args="-d $(find
turtlebot_follower)/follow.rviz"/>
 <!--modify end -->
</launch>
```

You can put this file in turtlebot/src/turtlebot_apps/turtlebot_follower/launch/myfollower to run, and start the robot and camera:

```
$ roslaunch turtlebot_follower follower.launch
```

The parameters can be modified as needed. For example, in practice, we have found that following parameters work better when they take the following values.

```
<param name="x_scale" value="5.0" />
<param name="z_scale" value="2.0" />
<param name="min_x" value="-0.25" />
<param name="max_x" value="0.25" />
<param name="min_y" value="0.1" />
<param name="max_y" value="0.5" />
<param name="max_z" value="1.4" />
<param name="goal_z" value="0.7" />
```

When start following, the person should walk in front of the robot and the robot should follow the person. Approaching the robot will cause it to move backwards, slowly to the left or right, and the robot will follow the person as it turns. The follow function can be stopped by quickly leaving the robot.

You can modify the function implementation source code as needed turtlebot/src/turtlebot_apps/turtlebot_follower/src/follower.cpp.

7.2 Implementation of the Robot Waving Recognition Function

With the rapid development of technology, robots appear more and more in life, and people are increasingly pursuing a more natural and convenient human-computer interaction experience. From the original command line interface to the current touch graphical interface, human-computer interaction has been developing in the direction of simplicity, humanization, friendliness and naturalness. In recent years, the field of human-computer interaction has continued to advance, with the emergence of face recognition, voice recognition, human motion recognition, gesture recognition and many other new interaction methods.

In order to achieve a more natural interaction between robots and humans, in addition to voice interaction, people want robots to communicate through "body language" like humans do. Among the body language, waving is the most common gesture. We often see this in our lives when a person greets another person from a distance, often by waving rather than calling out. Similarly, with advances in robotics, people can give commands by waving to a robot when it is at a distance from a person and the human needs a service from the robot. Especially in crowded situations, it is more appropriate to summon the robot to approach them by waving.

At present, although sound source localization can solve this problem to a certain extent, it is limited to localization in a certain direction, and when there are multiple people located in the same direction, the robot is unable to make a judgment. At this point, we can use waving to indicate to the robot that we need its service. If the robot has a waving hand detection function, it can effectively improve the convenience of human-robot interaction.

7.2.1 *Implementation Framework and Difficulties Analysis of Robot Hand Waving Recognition*

This book uses a method of waving hand recognition based on RGB image face recognition using the Primesense camera on the upper part of the robot. Its simple flow chart is shown in Fig. 7.3. First, the robot uses a face detection algorithm to detect faces and rejects faces that are detected incorrectly based on skin color, determines the approximate range of the hand position when waving based on the location information of the face, and uses template matching and skin color pixel scaling to determine the waving person. Then it locks the orientation of the waving person, keeps moving to the waving person, and determines whether it has reached the waving person based on the size of the face in the image, thus serving the waving person.

The main difficulties in robot waving recognition are as follows:

1) Generally, human hands occupy a small area in the image, so it is difficult to find and locate them in the image.
2) For the human hand, there is no stable feature to characterize it or to detect the position of the hand in a stable way.
3) Since the robot is in motion, it increases the difficulty of waving hand recognition.

To address the above difficulties, this book uses a face recognition-based approach to robot waving recognition, which has the following advantages.

1) Based on the fact that the human face composed of human eyes, nose and mouth has an inverted triangular structure, the recognition is relatively high, and the existing face detection technology is also more mature and has a high accuracy rate.
2) Use only image RGB information, no depth information, simple and effective processing, can be applied to general RGB cameras.
3) When a person waves his or her hand, the palm of the hand is located just to the sides of the face or slightly in front of the position, and the height is approximated, and the skin color of a single person's face and hand is approximated, making it easy to identify.

Fig. 7.3 Flow chart of
robot waving recognition

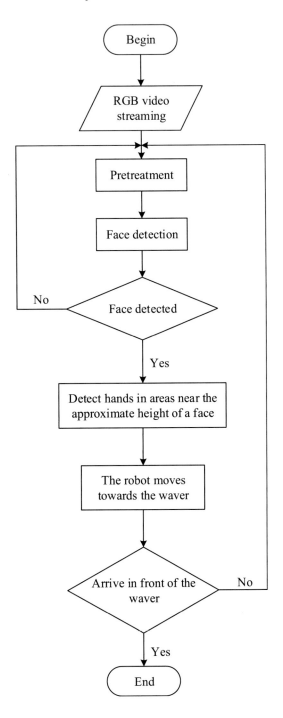

7.2.2   Face Detection Based on AdaBoost and Cascade Algorithms

In 2001, Viola and Jones pioneered the Haar rectangular feature and integral image approach for face detection based on the AdaBoost algorithm.In 2002, Rainer Lienhart and Jochen Maydt proposed extended Haar features, demonstrating that the new Haar feature set improves detection. The main points of the algorithm are as follows:

Image detection using Haar-like features.

Accelerating the computation of Haar-like features using integral images.

Strong classifier trained to perform face detection using AdaBoost algorithm.

Cascade the strong classifiers by using Cascade algorithm to screen the faces, so as to improve the accuracy.

(1) Haar rectangular feature

Using the Haar rectangular features of the input image, Lienhart et al. used three rectangular features: Haar-like boundary features, thin line features and center fatures, as shown in Fig. 7.4.

The purpose of these rectangular features is to quantify face features to distinguish between human faces and non-human faces. Place the rectangle in Fig. 7.4 over the face region and subtract the pixel sum of the black region from the pixel sum of the white region to obtain the face feature value. If this rectangle is placed over the non-face region, the calculated feature value should be different from the face feature value and the larger the difference, the better.

(2) Integral image

An integral image is a matrix representation capable of describing global information. It is constructed in such a way that the value at position I is the sum of all pixels in the direction of the upper left corner of the original image.

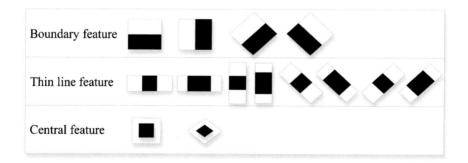

Fig. 7.4 Haar-like rectangular features used by Lienhart et al

Fig. 7.5 The result of pixel accumulation and operation of any matrix region in the image

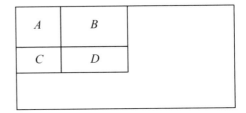

$$I(i, j) = \sum_{k \le i, l \le j} f(k, l) \tag{7.1}$$

The integral image construction algorithm is as follows;

1) Denote the cumulative sum in the row direction by $s(i,j)$ and initialize $s(i, -1) = 0$.
2) Denote an integral image by $I(i,j)$ and initialize $I(-1, i) = 0$.
3) Scan the image row by row and recursively compute the cumulative sum $s(i, j)$ and the value of the integral image $I(i,j)$ for each pixel (i,j) in the row direction.

$$s(i, j) = s(i, j - 1) + f(i, j) \tag{7.2}$$

$$I(i, j) = I(i - 1, j) + s(i, j) \tag{7.3}$$

Scanning the image once, the integral image I is constructed when the lower right pixel of the image is reached.

After constructing the integral image, the pixel sum of any matrix region in the image can be obtained by simple operations, as shown in `Fig. 7.5`.

Let the four vertices of D are $\alpha, \beta, \gamma, \delta$ and the pixel sum of D can be expressed as:

$$Dsum = I(\alpha) + I(\beta) - (I(\gamma) + I(\delta)) \tag{7.4}$$

It can be seen that the Haar-like feature value is the difference between the sum of two matrix pixels and can be done in the time of constant level.

(3) Flow of the Adaboost algorithm

We can compute multiple Haar-like rectangular features to get a feature value with greater discrimination. What the Adaboost algorithm needs to do is what kind of rectangular features to choose and how to combine them to make face detection better. The Adaboost algorithm is a classifier algorithm that was developed by Freund and Robert E. Schapire proposed in 1995. The theory proves that the error rate of a strong classifier will tend to zero when the number of simple classifiers tends to infinity, as long as each simple classifier has better classification ability than random guesses. The specific algorithm is as follows.

Given: $(x_1, y_1), \ldots, (x_m, y_m)$, among them $x_i \in X$, $y_i \in Y = \{-1, +1\}$
Initialization weights $D_1(i) = 1/m$,
For $t = 1, \ldots, T$:
Training weak classifiers with distributions D_t
Obtains a weak classifier: $h_t : X \rightarrow \{-1, +1\}$ its error is
$$\varepsilon_t = \Pr_{i \sim D_t}[h_t(x_i) \neq y_i] \quad (7.5)$$
Option:
$$\alpha_t = \frac{1}{2} \ln \left(\frac{1 - \varepsilon_t}{\varepsilon_t} \right) \quad (7.6)$$
Updated:
$$D_{t+1}(i) = \frac{D_t(i)}{Z_t} \times \begin{cases} e^{-\alpha t} & \text{if } h_t(x_i) = y_i \\ e^{\alpha t} & \text{if } h_t(x_i) \neq y_i \end{cases} = \frac{D_t(i) \exp(-\alpha_t y_i h_t(x_i))}{Z_t} \quad (7.7)$$
Z_t is the normalization factor(D_{t+1} is a distribution). Output the final strong classifier:
$$H(x) = sign \left(\sum_{t=1}^{T} \alpha_t h_t(x) \right) \quad (7.8)$$

(4) Cascade algorithm architecture

The basic idea of the Cascade (cascade) algorithm is to construct a multilayer cascade classifier, as shown in Fig. 7.6. This multilayer structure is similar to a decreasing decision tree. First, the image to be detected is divided into multiple sub-windows, and the sub-windows are detected by each layer of the cascade classifier. During the detection process, if the suspected face window is not filtered by each layer of classifier, subsequent processing is performed; if the currently detected sub-window is determined to be non-face at one level of the classifiers at each level, the detection of the current window is completed and the detection of the next sub-window is started. This multi-layer cascade classifier improves the detection rate and the speed of operation.

7.2.3 Identifying Human Hands With the Template Matching Algorithm

Template Matching is one of the representative methods in image recognition. It compares a number of feature vectors extracted from the image to be recognized with the corresponding feature vectors of the template, calculates the distance between the image and the template feature vectors, and determines the classification using the minimum distance method. Template matching usually requires a library of standard templates to be built in advance.

Two images are required to implement the template matching algorithm.

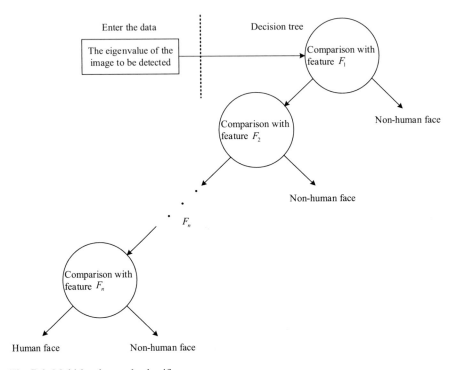

Fig. 7.6 Multi-level cascade classifier

Original image (I): in this image, the goal is to find a region that matches the template.

Sliding Template (T): the image block that will be compared to the original image.

To complete the region matching, the sliding template is compared with the original image. The sliding image block is moved one pixel at a time (in left-to-right, top-to-bottom order), while a metric is calculated to show whether the match is "good" or "bad" (or how similar the block image is to a particular region of the original image). For each position of T overlay I, the metric is stored in the resulting image matrix R, and each position (x, y) in R contains the match metric value.

The matching method used in this book is the squared difference matching method, i.e., method = CV_TM_SQDIFF, with the best matching time value of 0. The better the matching effect, the smaller its matching value, and R(x, y) is calculated as:

$$R(x, y) = \sum_{x',y'} (T(x\prime, \ y\prime) - I(x + x\prime, \ y + y\prime))^2 \qquad (7.9)$$

7.2.4 Skin Tone Segmentation Based on YCrCb Color Space

Skin tone segmentation based on the YCrCb color space is used in two places.

1) Rejecting the wrongly detected faces based on skin color.
2) Calculate the percentage of skin tone pixels in the region where the hand is located obtained by template matching.

The YCrCb color space is derived from the YUV color space, which is primarily used to optimize the transmission of color video signals so that black and white TVs can also represent color images through differences in grayscale. One of its advantages over RGB video signal transmission is that it occupies less bandwidth. YCrCb usually converts RGB images into YCrCb color space for image processing, where Y refers to the luminance or gray scale value, which is obtained by weighting the various parts of the RGB input signal together; U and V refer to the chrominance, which includes the two information of color hue and saturation are expressed by Cr and CB.respectively Cb represents the difference between the blue component and the luminance value in an RGB image; Cr represents the difference between the red component and the luminance value. They are relatively independent although both are chromaticity information of the image.

The general formula for converting RGB color space to YC_rC_B color space is:

$$\begin{cases} Y = 0.299R + 0.587G + 0.114B \\ C_r = R - Y \\ C_b = B - Y \end{cases} \tag{7.10}$$

The conversion formula for YUV color space to RGB color space is:

$$\begin{cases} R = Y + 1.14V \\ G = Y - 0.39U - 0.58V \\ B = Y + 2.03U \end{cases} \tag{7.11}$$

In the RGB color space, the R, G, and B color components contain luminance information and have some correlation. Therefore, the RGB color space is not well adapted for skin tone detection in terms of luminance. Many skin tone detection algorithms use a luminance-normalized RGB color space, but only the relative luminance components of the three colors are removed, in which luminance information still exists. Compared with other color formats, the YCrCb color format has the advantage of separating the luminance components from the colors, while the computational process is simple and the spatial coordinate representation is intuitive.

This book uses the CrCb plane of the YCrCb color space, and an input pixel is considered to belong to a skin tone pixel if its color falls into the region bounded by $= [140, 170]$ and $= [77 ,127]$.

7.2.5 Operational Testing of the Wave Recognition Function

Place 7.2.5wave_detect.cpp and 7.2.5wave_detect.launch from this book's resources in the /src and /launch folders of the package, respectively, and configure the image processing as described in Sect. 6.5.4.

> Note: You need to modify some file addresses in 7.2.5wave_detect.cpp to suit yourself, as well as placing the file haarcascade_frontalface_alt.xml in its corresponding location in the program. In Face_gender.cpp, the file is stored at /config, and also note where the resulting image is saved.

On the hardware side, connect Primesense to your computer.
Compile and run as follows:

```
$ cd robook_ws
$ catkin_make # compile before first run
$ source devel/setup.bash
$ roslaunch imgpcl wave_detect.launch
```

Since the procedure is only part of a comprehensive application scenario, initiating a wave recognition part can be achieved by sending a topic message and opening a new terminal:

```
$ rostopic pub detectWave std_msgs/String - wave
```

This is when a person can stand within visual range of the robot camera to perform the waving test.

7.3 Implementation of Object Recognition and Localization Functions of the Robot

7.3.1 Sliding Window Template Matching Method Based on Hue Histogram

This book uses a sliding window template matching method based on the Hue histogram for object recognition. Hue is a component of the HSV color space, which is often used in histograms, and its other two components are Saturation and Value, as shown in Fig. 7.7. Extracting the Hue channel as a color feature reduces the effect of uneven lighting conditions and improves recognition accuracy.

Hue
1
0
1
Saturation
Value
0

Fig. 7.7 The HSV color space and its three components

g1 g2 g3
g4 g0 g5
g6 g7 g8

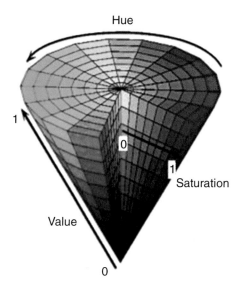

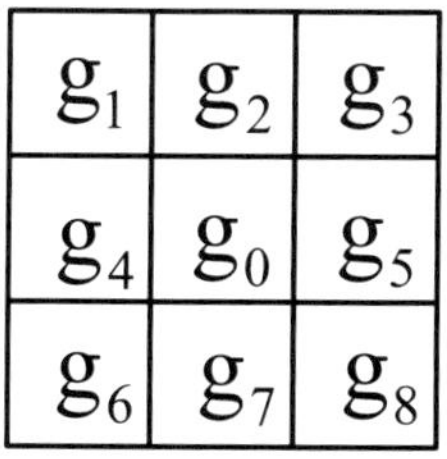

Fig. 7.8 LBP for 3 × 3 pixels

On the basis of sliding window, Hue histogram performs histogram matching of LBP (Local Binary Pattern) features. LBP features are a type of local binary features that focus on image texture. Fig. 7.8 show the LBP for 3 × 3 pixels.

Of which:

$$s(g_0, g_i) = \begin{cases} 1, & g_i \geq g_0, \\ 0, & g_i < g_0, \end{cases} \quad 1 \leq i \leq 8 \tag{7.12}$$

The LBP histogram distance of the candidate region from the template is:

$$LBP(g_0) = \sum_{i=1}^{8} s(g_0, g_1) \cdot 2^{i-1} \tag{7.13}$$

The specific steps are as follows:

1) Slide the window to search in the global image and calculate the Hue histogram distance between the current window and the template. If the similarity is greater

than the threshold s, it becomes a candidate region to go to step 2, otherwise it goes to step 3.

2) Calculate the LBP histogram distance between the candidate region and the template image, if the similarity is greater than the threshold s, it is considered as the target region, otherwise it is not.

3) Search for the next window.

Using the above sliding window template matching method based on hue histogram of color image, the region of object in two-dimensional image can be detected.

7.3.2 Object Localization Methods Based on Spatial Point Cloud Data

In addition to the task of object detection, the vision system has to provide the robot arm with the spatial coordinates of the object to be grasped. This book uses a depth camera to read the depth information of the object directly, i.e., spatial point cloud data.

Since the point cloud coordinates of a single point taken from the center of the object will be disturbed by various noises and cannot accurately locate the object, the point cloud data of all points in the recognition area are collected and these coordinate points are clustered using the Random Sampling Consensus Algorithm (RANSAC), which can remove boundary points and noise points. Points larger than half of the average distance are removed, and the remaining points are averaged to obtain the final coordinates as the material coordinates of the object. This process is shown in Fig. 7.9.

The random sampling consistency algorithm estimates the parameters of a mathematical model from a set of outlier data in an iterative fashion. The algorithm assumes that the data contains correct data and abnormal data (or called noise). The correct data is recorded as the inlier and the anomalous data is recorded as the outlier. Also, the RANSAC algorithm assumes that given a set of correct data, there exist model parameters that can be computed to satisfy these data. The basic implementation steps are as follows:

1) Randomly draw n sample data from the dataset and calculate the transformation matrix H, denoted as model M.

2) Calculate the projection error between all the data in the dataset and the model M. If the error is less than a threshold, add the inlier set I.

3) If the number of elements in the current inlier set I is greater than the optimal inlier set I_best, update I_best=I and update the number of iterations k.

4) If the number of iterations is greater than k, exit; otherwise, increase the number of iterations by 1 and repeat the above steps.

Since the point cloud data obtained by the camera is relative to the camera coordinate system, the object coordinate points need to be transformed into the robotic arm

Fig. 7.9 Get a flowchart of
the 3D coordinates of an
object

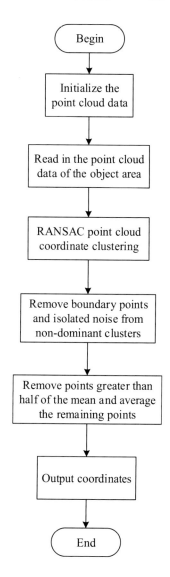

coordinate system in practice, so the homogeneous coordinate conversion matrix between the coordinate systems needs to be calculated based on the geometric relationship between the camera, the robotic arm and the object.

7.3.3 *Implementation and Testing of Object Recognition and Localization*

1. Capture templates
   In this example, the template is captured using the Primesense camera and the Primesense is connected to the computer. The following work is then performed.

   (1) Acquisition of fixed size templates
       Copy 7.3.3capimg.cpp and 7.3.3capimg.launch from this book's resources to the src and launch folders of the package, respectively, and configure CMakeLists.txt accordingly.
       The width of the capture template can be set in the book's resource 7.3.3capimg.cpp as follows.

```
int objWidth=50;
int objHeight=100;
```

   Modify the location where the capture template is saved as follows.

```
strs="/home/isi/robook_ws/src/imgpcl/template/";
```

   Compile and run:

```
$ cd robook_ws
$ catkin_make# to compile before first run
$ source devel/setup.bash
$ roslaunch imgpcl capimg.launch
```

   To capture a template, place the object in front of the camera so that the object is in the display box, then each time you click the "p" button, a template will be captured and saved.

   (2) Take a screenshot of the template with the mouse
       Copy 7.3.3capimg_mouse.cpp and 7.3.3capimg_mouse.launch from this book's resources to the src and launch folders of the package, respectively, and configure CMakeLists.txt accordingly.
       Set the location where the capture template is saved in capimg_mouse.cpp, as follows.

```
name="/home/isi/robook_ws/src/imgpcl/template/";
```

   Compile and run:

```
$ cd robook_ws
$ catkin_make # compile before first run
$ source devel/setup.bash
$ roslaunch imgpcl capimg_mouse.launch
```

You can then use the mouse to take a screenshot.

2. Change the name of the template image

Find the template image in /home/isi/robook_ws/src/imgpcl/template/, change the name afterwards and store it in /home/isi/robook_ws/src/imgpcl/template/.

Edit the names of the images in /home/isi/robook_ws/src/imgpcl/template/obj_list.txt. Click a carriage return for a line break between the different names and make sure the names in the obj_list.txt file are the same as the text of the images. In this example, the contents of obj_list.txt are as follows:

GreenTea.jpg
PotatoChips.jpg

3. Configuration program files

Store the file 7.3.3objDetect.cpp in the imgpcl/src folder, the file 7.3.3objDetect.launch in the imgpcl/launch folder, and the files objDetect.hpp, pos.h in the imgpcl/include folder in the resources of this book under the imgpcl/include folder, and modify the address variables involved in the files as you see fit. Also, configure CMakeLists.txt (refer to Sect. 3.2.8 for the creation of messages and services).

4. Compile and run

The laptop is connected to the Turtlebot robot and the Primesense camera, and the Primesense camera is used for object recognition and positioning, and the robot is gradually moved to the robot arm workspace by adjusting the robot's position in order to enable the robot arm to grasp the object.

```
$ cd robook_ws
$ catkin_make # compile before first run
$ source devel/setup.bash
$ roslaunch imgpcl objDetect.launch
```

Post the name of the target object to be identified via the topic.

```
$ rostopic pub objName std_msgs/String --potatoChips
```

7.4 Implementation of Face and Gender Recognition for Service Robots

As a service robot, it is especially important to recognize the owner. At the same time, it is important to prevent strangers from manipulating the robot and enhance its security. The intelligent service robot developed in this book is capable of recognizing faces through color images acquired by Primesense to distinguish the robot operator from strangers, thus confirming the owner and improving security. Face recognition in this book is implemented using the following two approaches: a traditional OpenCV-based face recognition and a Dlib library-based face recognition approach. The recognition of gender by the robot is achieved by the OpenCV-based gender recognition method.

7.4.1 Traditional Face and Gender Recognition Methods Based on OpenCV

The traditional OpenCV-based face recognition method consists of the following four main steps.

1) Face detection: the role is to locate the face region, here the concern is to identify the face. In this book the OpenCV trained Fast Haar detector haarcascade_frontalface_alt.xml is used.
2) Face pre-processing: the detected face images are adjusted and optimized, mainly using grayscale conversion and histogram equalization.
3) Collecting and training face models: in order to recognize faces, a sufficient number of face images to be recognized need to be collected. Using these collected face images, a model is trained online and saved. For each subsequent frame, the parameters in the model can then be matched online by the algorithm for recognition. The CV::Algorithm class provided by OpenCV is used in this book, with classes based on algorithms related to feature faces (PCA, Principal Component Analysis), Fisher faces (LDA, Linear Discriminant Analysis) and LPBH (Local Binary Pattern Histogram), which are created via cv::Algorithm:: create< FaceRecognizer> to create a FaceRecognizer object. After the FaceRecognizer object is created, the collected face data and labels are passed to the FaceRecognizer::train() function, and the model is trained. A similar approach can be used to train the robot to achieve male and female gender recognition.
4) Face recognition: calculate which face in the database is most similar to the current face. In this book, we use OpenCV's FaceRecognizer class and call the FaceRecognizer::predict() function to complete the face recognition.

The robot developed in this book is capable of recognizing the gender of a person from the color images acquired by Primesense. The recognition of gender is similar to the process of face recognition described above. It consists of four main steps as follows.

1) Collect samples and train the model: collect enough face images of men and women, train the model using the FaceRecognizer::train() function provided by OpenCV, and save the model.

2) Locate the face region: after detecting the face, use the OpenCV trained Fast Haar detector haarcascade_frontalface_alt.xml.

3) Face pre-processing: the face detected images are adjusted and optimized, mainly the processing of grayscale conversion and histogram equalization.

4) Online gender recognition: the parameters in the model are matched and recognized online by the algorithm, this book uses the FaceRecognizer::predict() function for face recognition.

7.4.2   Operational Testing of the OpenCV-Based Face and Gender Recognition Function

The OpenCV based face recognition and gender recognition algorithms need to be trained, where face recognition is done by collecting samples online (positive samples for the host's face and negative samples for other faces) and training, and gender recognition is trained offline, with positive and negative samples for male and female face photos, respectively.

1. Gender recognition training

The training for gender recognition is performed first. Find 7.4.2gender_train. cpp in the book's resources, place it under the imgpcl/src folder, and configure CmakeLists.txt for this. Place the training sample folder /7.4.2gender_train_img in the resources under /imgpcl/template/, place the index list 7.4.2gender_index. txt under /imgpcl/template/, and ensure that the 7.4.2gender_index.txt file sample address matches the actual sample address.

Note: Depending on the computer used, modify the variables in the 7.4.2gender_train.cpp file regarding addresses, including haarcascade_frontalface_alt.xml, operator_index.txt, eigenfacepca.yml, and picToPdf.sh. Among them, the picToPdf.sh script file is used to save the result image to a pdf document for easy viewing of the results later.

Compile and run:

```
$ cd robook_ws
$ catkin_make# to compile before first run
$ source devel/setup.bash
$ rosrun imgpcl gender_train
```

When run successfully, the eigenfacepca.yml file will be generated in the / imgpcl/config folder.

2. Test for identifying users and gender in the population

The main objective of this session is to perform OpenCV based face and gender recognition. First, the user stands in front of the robot and the robot remembers the user's look and gender. After that, the user comes to the crowd (around 7 people) and the robot should be able to find the user in the crowd and identify the gender of others.

The hardware environment required for this example is to connect the computer to the vision camera Primesense, the robot Turtlebot, and optionally to the speaker, which allowing the robot to speak the recognition results by voice.

Find 7.4.2face_gender.cpp in the book's resources and place it under the imgpcl/src folder with the relevant configuration for CmakeLists.txt; find the 7.4.2face_gender.launch file and place it under the imgpcl/src folder; place the training samples folder in the resources /7.4. 2gender_train_img under / imgpcl; place the index list 7.4.2gender_index.txt under /imgpcl and ensure that the sample addresses in the 7.4.2gender_index.txt file match the actual sample addresses.

> Note: The file address parameters involved in face_gender.cpp have to be modified depending on the computer being used.

Refer to Chapter 10 for ways to get the robot to output recognition results in speech. Place 7.4.2 person_recog.py and 7.4.2 person_recog.launch from the book's resources in the speech /src/ and speech/src/ folders, respectively, and also give execution rights to person_recog.py:

```
$ chmod +x person_recog.py
```

Run the following command:

```
$ cd robook_ws
$ catkin_make# to compile before first run
$ source devel/setup.bash
$ roslaunch imgpcl face_gender.launch
```

Open a new terminal and run the following command:

```
$ cd robook_ws
$ source devel/setup.bash
$ roslaunch speech person_recog.launch
```

7.4.3 Face Recognition Method Based on Dlib Library

Dlib is a C++ open-source third-party library containing machine learning algorithms and tools that have been used in a wide range of industrial and academic applications, including robotics, embedded devices, mobile phones, and large high-performance computing environments.

Dlib enables face detection and recognition and its algorithm implementation uses HOG features and cascading classifiers. The implementation of the algorithm is as follows.

1) Grayscale the image.
2) Use Gamma to correct the color space of the normalized image.
3) Calculate the gradient of each image pixel.
4) Divide the image into small cells.
5) Calculate the gradient histogram for each cell.
6) Combine small cells into larger blocks and normalize the gradient histogram in the blocks.
7) Generate the HOG feature description vector.

For face recognition methods based on the Dlib library, very few training samples are required, and only one training sample can be used. Because Dlib uses a structural support vector machine based training method, the method is able to train on all sub-windows of each image, which means that very time consuming sampling and "hard case mining" is no longer needed. Face recognition methods based on the Dlib library do not require many training samples to achieve good results, whereas OpenCV-based training methods require dozens of positive and negative samples, or even more.

7.4.4   Operational Testing of the Face Recognition Function Based on the Dlib Library

In this section, we will introduce a powerful, simple and easy to use face recognition open-source project face_recognition, based on this succinct face recognition library, we are able to use Python and command line tools to extract and recognize faces and perform operations on faces.

1. Install face_recognition

In a Python2 environment, install face_recognition using the following command:

```
$ pip install face_recognition
```

In a Python3 environment, install face_recognition using the following command:

```
$ pip3 install face_recognition
```

2. Install OpenCV

In a Python2 environment, install OpenCV using the following command:

```
$ pip install opencv-python
```

In a Python3 environment, install OpenCV using the following command:

```
$ pip3 install opencv-python
```

3. Install the PIL library
   In a Python2 environment, install the PIL library using the following command:

```
$ pip install Pillow
```

In a Python 3 environment, install the PIL library using the following command:

```
$pip3 install Pillow
```

4. Face detection procedure
   The face detection program face_detection.py used in this example is as follows:

```python
# coding=utf-8
from PIL import Image
import face_recognition
import cv2

#LoadImage
# Load the jpg file into a numpy array
image = face_recognition.load_image_file("all.jpg")


# Find all the faces in the image using the default HOG-based model.
# This method is fairly accurate, but not as accurate as the CNN model
and not GPU accelerated.
# See also: find_faces_in_picture_cnn.py


#Detecting the position of a face
#HOG model
face_locations = face_recognition.face_locations(image)
#cnn model
#face_locations = face_recognition.face_locations (image,
number_of_times_to_upsample=0, model="cnn")
print("I found {} face(s) in this photograph.".format(len
(face_locations)))
# Separate the detected faces and output the detected face images
for face_location in face_locations:

#Print the location of each face in this image
top, right, bottom, left = face_location
```

```
print ("A face is located at pixel location Top: {}, Left: {}, Bottom:
{}, Right: {}".format(top, left, bottom, right))
  # This code is used to cut out the recognized face
  # You can access the actual face itself like this:
  face_image = image[top:bottom, left:right]
  pil_image = Image.fromarray(face_image)
  pil_image.show()
```

We create a new folder named face as a face library, save the cut-out faces as pictures, and name the pictures with the corresponding person names. For example, we save the face picture of Xiao Ming to the face folder and name it xiaoming.jpg. Then, run the online face recognition program, when Xiao Ming's face appears in the camera's field of view, the face will be detected and recognized as "xiaoming.jpg". The face recognition program calculates the face features in the face database and matches them with the detected face features. When the similarity is greater than a certain threshold, it identifies it as the corresponding person in the face database.

5. Using Dlib in ROS

The following program receives the video stream from the Primesense camera by subscribing to a topic and performs the processing of face recognition, and then sends the names of the recognized faces as topics. The contents of the program dlibFace.py are as follows:

```
#! /user/bin/env python
# Need to be corrected
# coding=utf-8
from __future__ import print_function

import roslib
roslib.load_manifest('imgpcl')
import sys
import rospy
import cv2
from std_msgs.msg import String
from sensor_msgs.msg import Image
from cv_bridge import CvBridge, CvBridgeError
import face_recognition
import os
import time

# Data minimum position
def getMinIndex(my_list):
    min = my_list[0]
    for i in my_list:
      if i < min:
          min = i
    return my_list.tolist().index(min)
```

```
#face folder path
FindPath = "/home/isi/robook_ws/src/imgpcl/face "#
FileNames = os.listdir(FindPath)

# Load face in folder and learn how to recognize
known_name = []
known_image = []
known_face_encoding = []
face_number=0
for file_name in FileNames:
    fullfilename = os.path.join(FindPath,file_name)
    #print face_number,fullfilename
    known_name.append(file_name)
    known_image.append(face_recognition.load_image_file
(fullfilename))
    known_face_encoding.append(face_recognition.face_encodings
(known_image[face_number])[0])
    face_number += 1


......
# Load a sample picture and learn how to recognize it.
obama_image = face_recognition.load_image_file("people_face/
yangyikang")
obama_face_encoding = face_recognition.face_encodings
(obama_image)[0]

print "Next"
print obama_face_encoding
......
print (known_name)
# Initialize some variables
face_locations = []
face_encodings = []
face_names = []
process_this_cv_image = True
count=0
last_name = []
ifmaster=0
class image_converter:

  def __init__(self):
      self.image_pub = rospy.Publisher("image_topic_2",Image)
      self.person_name_pub = rospy.Publisher("person_name",String)
      self.bridge = CvBridge()
      self.image_sub = rospy.Subscriber("/camera1/rgb/image_raw",
Image,self.callback)

  def callback(self,data):
      if ifmaster==0:
          try:
              cv_image = self.bridge.imgmsg_to_cv2(data, "bgr8")
          except CvBridgeError as e:
```

```
        print(e)
    #############
        # Resize cv_image of video to 1/4 size for faster face recognition
processing
            small_cv_image = cv2.resize(cv_image, (0, 0), fx=0.333,
fy=0.333)

                # Find all the faces and face encodings in the current
cv_image of video
            face_locations = face_recognition.face_locations
(small_cv_image)
            face_encodings = face_recognition.face_encodings
(small_cv_image, face_locations)

            face_names = []
            for face_encoding in face_encodings:
                    # See if the face is a match for the known face(s)
                match = face_recognition.face_distance
(known_face_encoding, face_encoding)
                print (match)
                it=getMinIndex(match)

                if match[it]>0.6:
                    name='uknown'
                else:
                    name = known_name[it]

                face_names.append(name[:-4])
                global last_name
                global count
                if last_name==name:
                    count=count+1
                else:
                    count=0
                last_name=name
                if count == 2:
                    self.person_name_pub.publish(name[:-4])

                    if name[:-4]=="jintianlei":
                        global ifmaster
                        ifmaster=1
                        cv2.destroyAllWindows()
            global process_this_cv_image
            process_this_cv_image = not process_this_cv_image

            # Display the results
            for (top, right, bottom, left), name in zip(face_locations,
face_names):
            # Scale back up face locations since the cv_image we detected in
was scaled to 1/4 size
                top *= 3
                right *= 3
```

```
            bottom *= 3
            left *= 3

        # Draw a box around the face
        cv2.rectangle(cv_image, (left, top), (right, bottom), (0, 0,
255), 2)

        # Draw a label with a name below the face
        #cv2.rectangle(cv_image, (left, bottom - 35), (right, bottom),
(0, 0, 255), cv2.FILLED)
        cv2.rectangle(cv_image, (left, bottom - 35), (right, bottom),
(0, 0, 255), 2)
        font = cv2.FONT_HERSHEY_DUPLEX
        cv2.putText(cv_image, name, (left + 6, bottom - 6), font, 1.0,
(255, 255, 255), 1)

        # Display the resulting image
        cv2.imshow('Video', cv_image)

        # Hit 'q' on the keyboard to quit!
        #if cv2.waitKey(1) & 0xFF == ord('q'):
        #break

        cv2.waitKey(3)
    ##################
        try:
            self.image_pub.publish(self.bridge.cv2_to_imgmsg
(cv_image, "bgr8"))
        except CvBridgeError as e:
            print(e)

def main(args):
    ic = image_converter()
    rospy.init_node('image_converter', anonymous=True)
    try:
        rospy.spin()
    except KeyboardInterrupt:
        print("Shutting down")
    cv2.destroyAllWindows()

if __name__ == '__main__':
    main(sys.argv)
```

The reader can also find 7.4.4dlibFace.py in the book's resources and place it in the imgpcl/src/ folder with permissions; find 7.4.4dlibFace.launch in the imgpcl/launch/ folder; and place a face image in the imgpcl/face/ folder, taking care to modify the path location of the image in dlibFace.py. Use Primesense to complete the face recognition. The program recognizes the face through the Primesense camera according to the template and sends the name of the recognized face as a topic.

Run the following procedure to complete the above:

```
$ cd robook_ws
$ source devel/setup.bash
$ roslaunch imgpcl dlibFace.launch
```

7.5   Using TensorFlow to Recognize Handwritten Numbers

7.5.1   Introduction to TensorFlow

TensorFlow is open-source software that uses a data flow graph to perform numerical computations, and its flexible architecture supports users to perform computations on multiple platforms, such as one or more CPUs (or GPUs), servers, mobile devices, etc. TensorFlow was originally developed by the Google Brain team (part of Google's machine intelligence research agency) and researchers working on machine learning and deep neural network research. Due to the versatility of the system, TensorFlow has been widely used in other areas of computing as well.

Some of the learning resources for TensorFlow are given below.

Official TensorFlow website in English: http://tensorflow.org/

TensorFlow Chinese community: http://www.tensorfly.cn/

Official GitHub: https://github.com/tensorflow/tensorflow

Chinese version of GitHub: https://github.com/jikexueyuanwiki/tensorflow-zh

7.5.2   Installing TensorFlow

On Ubuntu 14.04, the easiest way to install TensorFlow is with pip. pip is a tool for installing and managing Python packages. pip allows you to install TensorFlow already packaged and the dependencies TensorFlow needs.

First, determine the version of Python you have installed on your computer. Enter the following command in the terminal.

```
$ python
```

The terminal will then display the version of Python that already exists.

Then, install the pip command:

```
$ sudo apt-get install python-pip python-dev # for Python 2.7
$ sudo apt-get install python3-pip python3-dev # for Python 3.n
```

Next, install TensorFlow:

```
$ sudo pip install tensorflow # Python 2.7; CPU support
$ sudo pip3 install tensorflow # Python 3.n; CPU support
```

> Note: As each computer is different, different errors may occur during the installation process, you can search for a solution based on the errors that occur.
>
> There is more than one way to install TensorFlow, there is also installation via Anaconda, Docker based installation and other methods that you can get the latest installation wizard from https://www.tensorflow.org/install/install_linux.
>
> TensorFlow is available in CPU and GPU versions, the installation of the CPU version is given above. Nvidia graphics drivers, CUDA, CuDNN need to be installed before installing the GPU version.

Next, let's briefly test if the TensorFlow installation is successful. Open a terminal, type "Python", and execute the following command line to see if the output works.

```
import tensorflow as tf
sess = tf.Session()
hello=tf.constant('Hello, Tensorflow!')
print(sess.run(hello))
a = tf.constant(28)
b = tf.constant(47)
print sess.run(a+b)
```

The terminal displays as shown in Fig. 7.10:

7.5.3   Basic Concepts of TensorFlow

TensorFlow is characterized by the use of graphs to represent computational tasks; execution of graphs in contexts called sessions; use of tensors to represent data; maintenance of state through variables; and the use of provisioning and extraction to copy data for or from arbitrary operations.

1. Graph

    TensorFlow uses graphs to represent computational tasks. The nodes in the graph are called op (operation). An op obtains 0 or more tensors and performs a computation that yields 0 or more tensors, each of which is a typed multidimensional array. For example, a set of image sets can be represented as

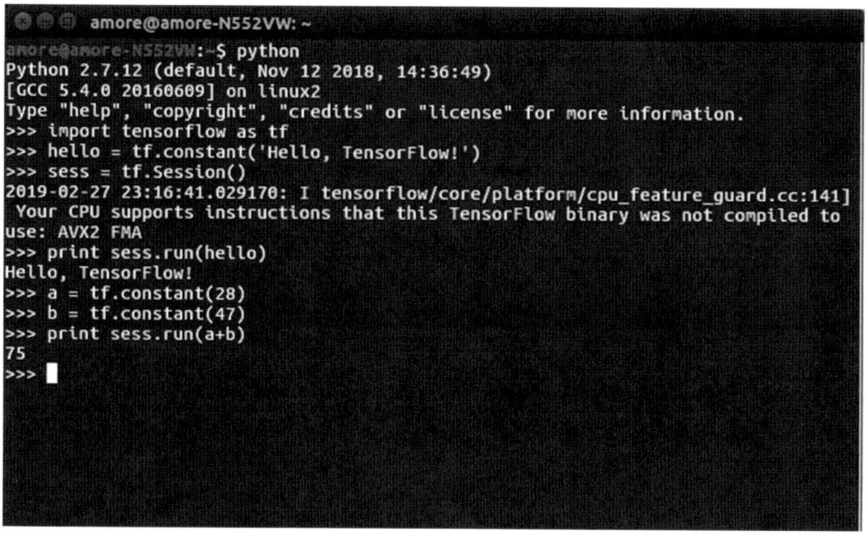

Fig. 7.10 TensorFlow installation success screen

a four-dimensional array of floating-point numbers [batch, height, width, channels].

For the test code in the previous section:

```
hello=tf.constant('Hello, Tensorflow!')
a = tf.constant(28)
b = tf.constant(47)
```

One of the operations performed on the graph is to use the tf.constant() method to create a constant op which will be added to the graph as a node, this is what constructs the graph.

2. Sessions

After the build phase is complete, the graph needs to be started in a session. First, create a Session object.

```
sess = tf.Session()
```

The Session class will place all operations or nodes on a computing device like a CPU or GPU. If there are no creation parameters, the session constructor will start the default graph.

The function call run() triggers the execution of op in the graph:

```
print(sess.run(hello))
```

This means that it will execute an op named "hello" and print it out in the terminal.

Session objects need to be closed after use to release resources, and can be closed automatically using the "with" block. For example:

```
with tf.Session() as sess:
    result = sess.run(hello)
    print result
```

3. Tensor

The tensor data structure represents all data, and all the data passed between operations is tensor. Tensor can be thought of as an n-dimensional array or a list, and a tensor contains a static type rank and a shape.

4. Variables

Variable maintains information about the state of the graph during execution. During runtime, if you need to save the state of an operation, you can use tf. Variable() to do so. The following example demonstrates how to use tf.Variable ().

Create a variable, initialized to scalar 0:

```
state = tf.Variable(0, name="counter")
Create an op that acts to add 1 to state (tf.assign acts to assign the
value of new_value to state).
one = tf.constant(1)
new_value = tf.add(state, one)
update = tf.assign(state, new_value)
After starting the graph, you need to use tf.initialize_ all_ The
variables() function initializes a variable once: init_op = tf.
initialize_all_variables()
```

Run the diagram to make it effective:

```
with tf.Session() as sess:
sess.run(init_op)
print sess.run(state)
for _ in range(3):
sess.run(update)
print sess.run(state)
```

This enables the counting function.

5. Extraction

After the session has finished running, we can view the results of the session run by fetch. When using the run() method in the Session object, the op is passed to run() and the output is fetched as a Tensor. Example:

```
a = tf.constant(28)
b = tf.constant(47)
```

```
add = tf.add(a,b)
sess = tf.Sessions()
result = sess.run(add)
print(result)
```

The extraction can be from a single or multiple Tensor. It is worth noting that if multiple Tensor values need to be obtained, they are obtained together in a single run of the op, not one by one.

6. Supply

The feed mechanism allows to feed the Tensor during graph execution. At this point, you first need to use tf.Placeholder () function to create placeholders and define the feed object, after which the feed object can be used as an argument to the run() method call.

```
x = tf.placeholder(tf.float32)
y = tf.placeholder(tf.float32)
output = tf.multiply(x,y)
with tf.Session() as sess:
    print(sess.run([output],feed_dict={x:[7.],y:[2.]}))

# output:
# [array([14.], dtype=float32)]
```

Note that the feed object is only valid within the method that calls it; when the method ends and the feed disappears.

7.5.4 Handwritten Digit Recognition Using TensorFlow

The example project package for this section is ch7_ros_tensorflow and the main program is example_mnist.py. The mnist dataset for this program is from the National Institute of Standards and Technology and is available from http://yann.lecun.com/exdb/mnist/ or can be imported into the project via the project package at input _data.py is imported into the project, and the model is obtained by training it with train.py. Next, the main code of the example_mnist.py program will be analyzed paragraph by paragraph.

First, the module needs to be imported:

```
import rospy
from sensor_msgs.msg import Image
from std_msgs.msg import Int16
from cv_bridge import CvBridge
import cv2
import numpy as np
```

```
import tensorflow as tf
import os
```

Among them, rospy has the ROS Python API; imports Image messages from sensor_msgs and processes image messages; cv_bridge implements conversions between ROS images and OpenCV data types; numpy and TensorFlow modules are used to classify images; and system commands are called using the os module.

After constructing the CNN convolutional neural network model, the second half of the code is the constructor for the class RosTensorFlow():

```
class RosTensorFlow():
  def __init__(self):
```

Create a cv_bridge object for ROS and OpenCV image conversion:

```
self._cv_bridge = CvBridge()
```

Save the model using tf.train.Saver(), create the Session() object. Initialize the variables and run the session:

```
self._saver = tf.train.Saver()
self._session = tf.InteractiveSession()
init_op = tf.initialize_all_variables()
self._session.run(init_op)
```

Use saver.restore() to read the model from the specified path:

```
ROOT_PATH = os.path.abspath(os.path.join(os.path.dirname(__file__),
os.pardir))
PATH_TO_CKPT = ROOT_PATH + '/include/model.ckpt'
self._saver.restore(self._session, PATH_TO_CKPT)
This application subscribes to the topic "/image_raw" and publishes the
results of the recognition in "/result". The subscriber and publisher
handles are as follows.
self._sub = rospy.Subscriber('usb_cam/image_raw', Image, self.
callback, queue_size=1)
self._pub = rospy.Publisher('/result', Int16, queue_size=1)
```

Convert ROS image messages to image callbacks of the OpenCV data type and perform some other processing:

```
def callback(self, image_msg):
    cv_image = self._cv_bridge.imgmsg_to_cv2(image_msg, "bgr8")
    cv_image_gray = cv2.cvtColor(cv_image, cv2.COLOR_RGB2GRAY)
    ret,cv_image_binary = cv2.threshold(cv_image_gray,128,255,cv2.
THRESH_BINARY_INV)
    cv_image_28 = cv2.resize(cv_image_binary,(28,28))
    np_image = np.reshape(cv_image_28,(1,28,28,1))
```

TensorFlow does the identification and puts the possible results into the predict_num array. Then, the most probable values are taken and the results are finally obtained and published under the "/result" topic:

```
predict_num = self._session.run(self.y_conv,feed_dict = {self.x :
np_image, self.keep_prob : 1.0})
answer = np.argmax(predict_num,1)
rospy.loginfo('%d' % answer)
self._pub.publish(answer)
```

Initialize the class and call the main() method in the RosTensorFlow() object:

```
def main(self):
   rospy.spin()
if __name__ == '__main__':
   rospy.init_node('ros_tensorflow_mnist')
   tensor = RosTensorFlow()
   tensor.main()
```

The main() method will run the node's spin() and execute a callback whenever the theme "/image_raw" has image data coming in.

Create the example_mnist.laugh file under the launch file:

```
<launch>

<node pkg = "ch7_ros_tensorflow" name = "example_mnist" type =
"example_mnist.py" output = "screen">
 </node>

</launch>
```

Next, the nodes were run for handwritten digit recognition. Running one of the nodes successfully assumes that the OpenCV and usb_cam packages are installed.

Running roscore:

```
$ roscore
```

Running camera:

```
$ roslaunch usb_cam usb_cam-test.launch
```

Running the handwritten digit recognition node:

```
$ roslaunch ch7_ros_tensorflow example_mnist.launch
```

View the results of the /result topic via the echo command:

```
$ rostopic echo /result
```

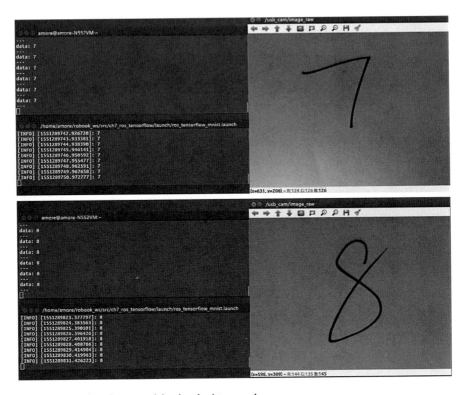

Fig. 7.11 Interface for recognizing handwritten numbers

The screen for performing recognition is shown in Fig. 7.11.

This chapter focuses on the functions of the robot to recognize and follow its owner, the function to recognize the wave call of the owner from multiple people, the object recognition and localization function, the function to recognize the face and gender within the field of view, and the function to recognize handwritten numbers using TensorFlow. All these functions are implemented based on the robot vision function. The advanced learning of robot vision functions helps to implement more robots with more operational functions and applications.

Exercises

1. Implementing the use of voice commands to instruct the robot to follow and stop following with voice commands.

2. Waving a hand to summon a robot in an environment with more than 3 people

3. Recognizes faces and owners in environments with more than 4 people, determines gender and announces it by voice.

Further Reading

1. Liu Song. Research on partial special pose recognition and cloud application extension for mobile service robots [D]. Hefei: University of Science and Technology of China, 2016.
2. Viola P, Jones M. Rapid Object Detection using a Boosted Cascade of Simple Features [C]. Proceedings of the 2001 IEEE Computer Society Conference on Computer Vision and Pattern Recognition, Kauai, HI, USA, Dec. 8-14, 2001, 511-518.
3. Wang Y. Application of AdaBoost algorithm in face detection [J]. Computer Fans, 2017(4), 192-192.
4. Freund Y, Schapire R E. A desicion-theoretic generalization of on-line learning and an application to boosting [C]. European Conference on Computational Learning Theory. 1995: 119-139.
5. Linlin Wang. Research on face detection based on skin color model and AdaBoost algorithm [D]. Xi'an: Chang'an University, 2014.
6. Li Jing. Design of intelligent recognition system for traffic signs based on TensorFlow [D]. Tianjin: Tianjin University of Technology, 2018.
7. Qiu D. Research on skin color detection based on HSV and YCrCb color space [J]. Computer Programming Skills and Maintenance, 2012(10):74-75.
8. Wu, Gao-Ling. Design and implementation of face detection algorithm based on YCrCb color space [D]. Chengdu: University of Electronic Science and Technology, 2013.
9. Zhang Jingzhen, Shi Yuexiang. Skin color clustering face detection method in YCgCr color space [J]. Computer Engineering and Applications, 2009, 45(22):163-165.
10. Qin Yuefu, Zhi Hang, Xu Yi. Improved Camshift target tracking algorithm based on 3D histogram[J]. Modern Electronics Technology, 2014, 37(2): 29-33, 37.
11. Dlib. Dlib C++ Library [EB/OL]. http://dlib.net/.
12. Zhang Zhi-Ling. Python implementation of deep learning-based face recognition [J]. E-commerce, 2018(5): 47, 96.
13. Yue Y. Design and implementation of a depth camera-based face and pedestrian perception system [D]. Hangzhou: Zhejiang University, 2017.
14. Binjue Zheng. Kinect depth information for gesture recognition [D]. Hangzhou: Hangzhou University of Electronic Science and Technology, 2014.

Chapter 8
Autonomous Robot Navigation Function

Robots often need to move autonomously in a given space to respond to calls and deliver objects, which requires autonomous positioning and navigation capabilities. To achieve navigation functions, having a mobile base and a perceptron to collect environmental information is necessary. This book uses the robot TurtleBot2, i.e., a Kobuki base for development, and robot collects environmental information through its vision sensor, Kinect Xbox 360, in order to achieve navigation.

In this chapter, we will first introduce key techniques for navigation, including robot localization and map building, and path planning. Then we will introduce the kinematic analysis of the Kobuki base model. Finally, we will introduce the navigation stack for ROS and learn how to configure and use the navigation stack on Turtlebot.

Through this chapter, the reader should gain an initial understanding of the critical navigation, understand the kinematic model of the robot base, and learn how to implement the navigation motion of the TurtleBot2 robot using the ROS navigation stack.

8.1 Key Technologies for Autonomous Robot Navigation

The so-called robot navigation refers to the robot sensing the environment and its state by its own sensors or sensors in the environment, drawing a map, and planning a global path according to the map information, according to the target point given by the system. In the process of moving, it constantly senses the change of environmental information, carries out local path planning, achieves collision-free autonomous movement in the environment with obstacles, and finally reaches the given target point to complete the given task.

Environment perception, navigation, and localization are the basic problems for mobile robots to achieve autonomy and intelligence.

© The Author(s), under exclusive license to Springer Nature Singapore Pte Ltd. 2023
F. Duan et al., *Intelligent Robot*, https://doi.org/10.1007/978-981-19-8253-8_8

8.1.1 Robot Localization and Map Building

8.1.1.1 Method of Robot Localization

Robot localization is the process of determining the pose (position and heading) of the robot in the two-dimensional or three-dimensional environment in which it is located, and different localization methods require different sensor systems and processing methods. The two main positioning methods commonly used are relative positioning and absolute positioning.

The two main methods of relative localization are as follows:

1. Dead reckoning: With the current position known, the next position can be projected by measuring the distance and bearing moved. Odometry and gyroscope are the sensors often used for dead reckoning. The advantages of this method are the high sampling rate, simple calculation, and low price. However, this method is prone to the accumulation of positioning errors and must be corrected in some way on an ongoing basis.
2. Inertial localization: By using gyroscopes to measure angular velocities and accelerometers to measure accelerations, the distances traveled and the angles of deflection relative to the starting position can be calculated based on the primary and secondary integrations of the measured values.

The basic principle of relative positioning is to calculate the robot's distance value and deflection angle relative to the initial position from the sensor measurements. As time increases, errors are usually accumulated and magnified, making the relative positioning method unsuitable for moving over long distances. In contrast, absolute positioning can effectively reduce the accumulated error because it achieves positioning by measuring the absolute position of the robot. The main methods of absolute positioning are as follows.

1. Landmark localization: This localization is achieved by measuring the relative position between the robot and the landmark localization. The road signs can be manually placed road signs or natural road signs with distinctive features that can be easily recognized.
2. Satellite localization: It refers to locating robot by high-precision space satellite. Satellite localization has a wide range, but suffers from the problem of large deviation in positioning at close range.
3. Matching localization: This localization is constructing a local map from the environmental information acquired by the robot's own sensors and then comparing it with the complete global map to calculate the robot's current position and heading. This method is suitable for moving in a known environment.

The navigation engineering stack in ROS uses a matched localization approach in navigation with global maps, which is localized by the KLD-AMCL algorithm. The AMCL (Adaptive Monte Carlo Localization) method is a probability-based localization system for mobile robots in a two-dimensional environment and uses particle filter to position the robot in the known maps for positional tracking.The ROS navigation engineering stack uses the KLD (Kullback Leibler Distance) sampling method, the KLD-AMCL algorithm, which reduces the number of particles and ensures a lower bound on the sampling error by computing the posterior probability of the maximum likelihood estimated sample and KL (Kullback-Leibler) divergence.

KLD-AMCL assumes that the particles follow a discrete piecewise-constant distribution that can be represented by k distinct bins of $X = (X_1, \ldots, X_k)$ and denotes the probability of each bin by $p = (p_1, \ldots, p_k)$. The processing flow of the KLD-AMCL algorithm is as follows:

Inputs:$S_{t-1} = \left\{ \left\langle x_{t-1}^{(i)},\ \omega_{t-1}^{(i)} \right\rangle i = 1,\ \ldots,\ n \right\}$ ($x_{t-1}^{(i)}$ is a state; $\omega_{t-1}^{(i)}$ is a non-negative numerical factor called importance weight; u_{t-1} is a control measure; z_t is observation, ε and δ are bounds and $n_{x_{\min}}$ is minimum sample size)

$S_t := 0,\ n = 0,\ n_x = 0,\ k = 0,\ \alpha = 0;$

Do

    Sample an index j from the discrete distribution given by the weights in S_{t-1};

    Use $x_{t-1}^{(j)}$ and u_{t-1} sampling $x_t^{(n)}$;

    $\omega_t^{(n)} := p\left(z_t \middle| x_t^{(n)} \right);$

    $\alpha := \alpha + \omega_t^{(n)};$

    $S_t := S_t \cup \left\{ \left\langle x_t^{(n)},\ \omega_t^{(n)} \right\rangle \right\};$

    if($x_t^{(n)}$ falls in empty bin b) **then**

        $k := k + 1$

        $b := non - empty$

        $n_x := \frac{k-1}{2\varepsilon} \left\{ 1 - \frac{2}{9(k-1)} + \sqrt{\frac{2}{9(k-1)}} z_{1-\delta} \right\}^3$

    $n := n + 1$

while $(n < n_x)$

for $i := 1, \ldots, n$ **do**

    $\omega_t^{(i)} := \omega_t^{(i)} / \alpha$

return S_t

8.1.1.2   Map Building Methods

Map building is the process of building a model of the environment in which the robot is located. In this process, a map for robot movement path planning is drawn by describing the robot's working environment and combining it with various sensor information. Commonly used map representations are occupancy grid maps, geometric maps, and topological maps. The ROS navigation project package set uses occupancy grid maps.

1. The occupancy grid map is based on the principle of dividing the entire working environment into a series of grids of the same size. Each grid is assigned a value of 0 or 1. 1 means the grid is occupied, and 0 means the raster is free. Also, a

threshold is set for each raster, and when the probability of the raster being occupied by an obstacle exceeds the threshold, the grid state is marked with a value of 1 and vice versa. Grid maps do not rely on the specific shape of the object, but are divided in a probabilistic manner and are therefore easy to create and maintain. However, the method is more limited to the size of the environment. When the working environment is large, the number of grids will increase, the memory occupied will also increase, and updating the maintained maps will take a lot of time, leading to reduced real-time performance.

2. Geometric map refers to the description of the environment by extracting relevant geometric features (e.g., points, lines, and surfaces) from the environmental information collected by the sensors. To obtain the geometric features of the environment, we also need a certain amount of sensor data, which has been processed accordingly. Also, to ensure the consistency of the map, the location of each observation must be required to be relatively accurate.

3. The topological map represents the important location points in the environment as nodes. The connecting lines between the nodes represent the path information between the important locations, and the weights represent the corresponding distance cost. This method represents a more compact map, occupies less space, does not require access to accurate location information, and therefore path planning is faster. The disadvantages of the method are that it is difficult to create and maintain, it tends to accumulate positioning errors, and it is not suitable for representing unstructured environments.

8.1.1.3 Simultaneous Positioning and Mapping

The Simultaneous Localization and Mapping (SLAM) problem of mobile robots is the basis for the navigation of mobile robots and one of the important conditions for their true autonomy and intelligence. SLAM is the integration of mobile robot localization and environment mapping, i.e., the robot builds an incremental environment map based on its own position estimation and sensor perception of the environment during its motion, and uses the map to realize its own localization.

The generic architecture of SLAM is shown in Fig. 8.1. The robot starts its motion with the departure position as the initial position, and since the robot knows nothing about the environment, it can only rely on its own odometer data for position estimation, while the environmental information obtained using sensors is processed accordingly and feature values are extracted to create a local map of the current position. The local map is updated by using the existing global map information for feature matching and updating the corresponding feature values with the observations to correlate the data. At the same time, the observations of the road signs are used to correct the current position of the robot and reduce the cumulative error due to the positional estimation.

The map is created using the slam_gmapping node of the gmapping package in the navigation stack of ROS, where a Rao-Blackwellized algorithm based on particle filtering is used to keep track of the robot's own position.

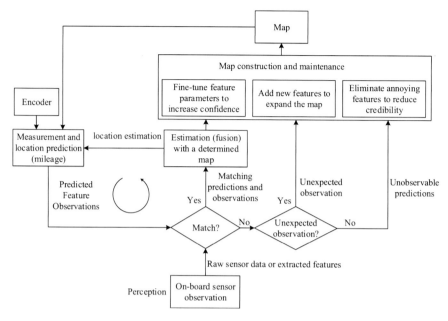

Fig. 8.1   General architecture of SLAM

8.1.2   Path Planning

The robot path planning problem can be described as finding a completely collision-free path for the robot to move from the start point to the goal point and travel as short as possible. According to the different requirements of the environment information during path planning, robot path planning algorithms can be divided into two categories: One is global path planning algorithms, which require that the entire environment information is known and cannot change during planning. The other is local path planning. In local path planning, the working environment in which the robot works is mostly dynamic and changing, and the robot has no prior knowledge of this environment.

8.1.2.1   Global Path Planning

The global path planning problem can be described as finding a collision-free optimal or suboptimal path in a known map of the environment. In the past 20 years, the optimal path planning problem has been a hot research topic in robot navigation technology. At present, many methods have been proposed by home and abroad researchers, which can be broadly classified into two categories: traditional methods and intelligent methods. Traditional methods are the commonly used A* algorithm, Dijkstra's algorithm, artificial potential field method, visual graph method, etc. In recent years, more and more intelligent methods have been applied to robotics, such as fuzzy control, neural networks and genetic algorithms.

The navigation project package set for ROS provides two global path planning algorithms, A* and Dijkstra. This section focuses on the A* algorithm.

The A* algorithm computes the minimum cost path on the cost map as the route for the global navigation of the robot. The valuation function of the A* algorithm is:

$$f(n) = g(n) + h(n)$$

where f(n) is the valuation function of the starting point through the current point n to the final target point; g(n) is the optimal surrogate value from the starting point to the current point n, which is deterministic; and h(n) is the estimated cost from the current point n to the target point.

The path search process for the A* algorithm is as follows.

1. Create two tables, OPEN and CLOSED, with all generated but unexamined points saved in the OPEN table and visited points saved in the CLOSED table. Put the initial point S into the OPEN table and set the CLOSED table to empty.
2. Repeat the following steps until you find the best path.
3. Traverse the OPEN table, find and use it as the best point and move this node into the CLOSED table.
4. If the current best point is the target point, a solution is obtained; if it is not the target point, the adjacent points of that point are used as successors and the following procedure is performed for each successor.

   (a) If the successor point is an obstacle or has been moved into the CLOSED table, this point is ignored.
   (b) If the successor point is not in the OPEN table, move it into the OPEN table and set the current best point as its parent node and record it.
   (c) If the successor point is in the OPEN table, we re-judging the points in the OPEN table. If there is a smaller value, the parent node of that successor point is set as the current best point and the sum is recalculated.
   (d) When the target point is moved into the OPEN table, it means the path is found. Otherwise the search fails and the OPEN table is empty, i.e., there is no path.

5. Save the obtained path which moves from the target point to the start point along with the parent node.

8.1.2.2 Local Path Planning-Robot Obstacle Avoidance

The commonly used robot obstacle avoidance algorithms are the APF (Artificial Potential Field) algorithm, VFH (Vector Field of Histogram) algorithm, DWA (Dynamic Window Approach) algorithm, etc.

1. APF algorithm

   The APF algorithm, the artificial potential field method, can be used not only for global path planning of the robot but also applied to real-time obstacle avoidance in local environments.

2. VFH algorithm

   The VFH algorithm is an obstacle avoidance algorithm proposed by Borenstein and Koren. The method uses a histogram in a two-dimensional Cartesian coordinate system to describe obstacle information, and each grid in the histogram possesses a deterministic value for the confidence that an obstacle is present in that grid.

3. DWA algorithm

   DWA is an algorithm derived from robot motion dynamics and is therefore particularly suitable for the operation of high-speed mobile robots. Unlike the previously described methods, DWA computes the control commands for robot translation and rotation directly in velocity space.

The ROS navigation stack set provides Dynamic Window Approaches (DWA) and Trajectory Rollout methods for local path planning.

The goal of local path planning is that given a tracking path and a cost map, the controller generates velocity commands and sends them to the movement base. By using the map, the planner creates a trajectory for the robot from the starting point to the target location. In the process, the planner at least creates a local value function at least around the robot, represented by a grid map. This value function encodes the cost of traversing the grid cells. The controller's job is to use this value function to determine the x, y, and theta velocities sent to the robot.

The basic idea of Trajectory Rollout is similar to that of the DWA algorithm, as follows.

1. Discrete sampling in the robot control space (dx, dy, dtheta).
2. For each sampling speed, forward simulations are performed from the current state of the robot to predict what will happen after a (short) period of time at the applied sampling speed.
3. Evaluate (score) each trajectory generated by the forward simulation using a metric containing the following characteristics: proximity to obstacle, proximity to the target, proximity to global path, and velocity, discarding illegal trajectories (those that collide with the obstacle).
4. Select the highest-scoring track and send the associated velocity to the mobile base.
5. Update and repeat the above steps.

The DWA differs from the Trajectory Rollout algorithm in how the robot control space is sampled. The Trajectory Rollout algorithm takes samples from the achievable velocities during the entire forward simulation and gives the robot's acceleration limits, while the DWA algorithm takes samples from the achievable velocities of only one simulation step and gives the robot's acceleration limit. This means that DWA is a more efficient algorithm because it samples in a smaller space, but since the DWA forward simulation is not a constant acceleration, the Trajectory Rollout algorithm may outperform the DWA algorithm for robots with lower acceleration limits. However, the practice has found that the DWA and Trajectory Rollout algorithms have similar performance, but using DWA improves efficiency.

8.2 Kinematic Analysis of the Kobuki Base Model

Motion control of the robot base is a prerequisite for navigation, and the design structure and drives of different types of bases determine different types of motion. Typical drive methods are differential drive, synchronous drive, and omni-directional drive. Since the Kobuki base used in this book has a two-wheel differential mechanical structure, only differential drives are described here.

The differential drive is a common drive method for autonomous mobile robots. In this approach, two symmetrical drive motors drive the robot forward or steer it by the differential speed ratio of the left and right wheels. The mechanical structure of the two-wheel differential robot base is shown in Fig. 8.2. The robot has two wheels of radius r, and both wheels are at a distance from the center of the robot; there are no steering wheels, only two coaxial fixed wheels, and steering is achieved by a two-wheel differential; adding one or more caster wheel does not change its kinematic characteristics. The Kobuki base has two symmetrical caster wheel in front and rear to increase its movement stability.

Fig. 8.2 also shows the global reference coordinate system $(X_G O_G Y_G)$ and the robot reference coordinate system $(X_R O_R Y_R)$ for the two-wheel differential robot. To determine the position of the robot, the center point O_R on the robot base is chosen as its position reference point. The position in the global reference coordinate system is represented by the coordinate sum, and the angular difference between the local

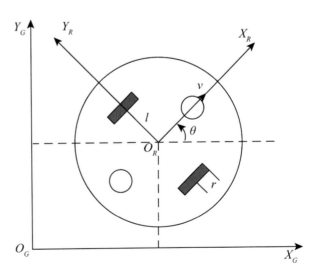

Fig. 8.2 Mechanical structure and kinematic model of the robot chassis

reference coordinate system and the global reference coordinate system is given by θ. The pose of the robot ξ_G can be described as a vector in terms of these three variables.

$$\xi_G = \begin{bmatrix} x \\ y \\ \theta \end{bmatrix} \tag{8.1}$$

The transformation between coordinate systems can be represented by the following standard orthogonal rotation matrix.

$$R(\theta) = \begin{bmatrix} \cos(\theta) & \sin(\theta) & 0 \\ -\sin(\theta) & \cos(\theta) & 0 \\ 0 & 0 & 1 \end{bmatrix} \tag{8.2}$$

The mapping relationship between coordinate systems is as follows.

$$\xi_R = R(\theta)\xi_G \tag{8.3}$$

The mapping of velocities between coordinate systems can be obtained as follows.

$$\dot{\xi}_R = R(\theta)\dot{\xi}_G \tag{8.4}$$

In robot trajectory planning, it is required to control the robot to reach a specified position at a specified velocity and attitude in each control step. If the velocity of each wheel is controlled directly without considering the dynamics of the system, the control is a kinematic control.

Given a wheel with a radius of r and the distance between the two wheels and the robot center O_R is l, the angular difference between the global and local reference coordinate systems is θ and the rotational speeds of the left and right wheels are ω_l and ω_r, the kinematic model of the robot in the world coordinate system is:

$$\dot{\xi}_G = \begin{bmatrix} \dot{x} \\ \dot{y} \\ \dot{\theta} \end{bmatrix} = f(l, r, \theta, \omega_l, \omega_r) \tag{8.5}$$

From the equation (8.4) for the mapping relationship between coordinate systems, it follows that

$$\dot{\xi}_G = R(\theta)^{-1}\dot{\xi}_R \tag{8.6}$$

The linear velocity of the left wheel is $r\omega_l$ and the linear velocity of the right wheel is $r\omega_r$, the linear velocity of the robot is the average velocity of the two wheels:

$$\dot{x}_R = r\frac{\omega_l + \omega_r}{2} \tag{8.7}$$

Because none of the wheels provide lateral motion in the robot reference coordinate system, so $\dot{y}_R$ is always 0.

If the left wheel is rotating alone and the robot's axis of rotation is around the right wheel, the angular velocity of rotation ω_1 can be calculated at the point O_R where the wheel instantaneously moves along the arc of a circle with radius $2l$.

$$\omega_1 = \frac{r\omega_l}{2l} \tag{8.8}$$

A separate transit of the right wheel can be calculated in the same way, except that the forward rotation produces a clockwise rotation at the point O_R.

$$\omega_2 = -\frac{r\omega_r}{2l} \tag{8.9}$$

The speed of the robot can be obtained as follows.

$$\dot{\theta}_R = r\frac{\omega_l - \omega_r}{2l} \tag{8.10}$$

Taking Eq. (8.7) and Eq. (8.10) into Eq. (8.6), we get:

$$\dot{\xi}_G = R(\theta)^{-1}\dot{\xi}_R = R(\theta)^{-1}\begin{bmatrix} r\dfrac{\omega_l + \omega_r}{2} \\ 0 \\ r\dfrac{\omega_l - \omega_r}{2l} \end{bmatrix} = \frac{r}{2}\begin{bmatrix} \cos(\theta) & -\sin(\theta) & 0 \\ \sin(\theta) & \cos(\theta) & 0 \\ 0 & 0 & 1 \end{bmatrix}$$

$$\times \begin{bmatrix} \omega_l + \omega_r \\ 0 \\ \dfrac{\omega_l - \omega_r}{l} \end{bmatrix} \tag{8.11}$$

Simplifying the above equation yields.

$$\dot{\xi}_G = \begin{bmatrix} \dot{x} \\ \dot{y} \\ \dot{\theta} \end{bmatrix} = \frac{r}{2}\begin{bmatrix} \cos(\theta) & 0 \\ \sin(\theta) & 0 \\ 0 & 1 \end{bmatrix}\begin{bmatrix} \omega_l + \omega_r \\ \dfrac{\omega_l - \omega_r}{l} \end{bmatrix} \tag{8.12}$$

Suppose the velocity of motion of the robot is $v = r\frac{\omega_l + \omega_r}{2}$, the angular velocity is $\omega = r\frac{\omega_l - \omega_r}{2l}$, the differential drive robot is modeled in the global coordinate system as

$$\dot{\xi}_G = \begin{bmatrix} \dot{x} \\ \dot{y} \\ \dot{\theta} \end{bmatrix} = \begin{bmatrix} \cos(\theta) & 0 \\ \sin(\theta) & 0 \\ 0 & 1 \end{bmatrix} \begin{bmatrix} v \\ \omega \end{bmatrix} \tag{8.13}$$

8.3 Navigation Package Set

8.3.1 Overview of the Navigation Package Set

The concept of a navigation engineering package set is fairly simple; it receives information from the odometer and sensor streams and outputs velocity commands to the mobile base. However, using the navigation stack on an arbitrary robot is a bit more complicated. The prerequisites for the use of the navigation engineering package set are that the robot must be running ROS, have an appropriate tf transformation tree, and publish sensor data using the correct ROS message type. In addition, the navigation engineering package set requires configuration of the robot's shape and dynamics parameters for high-level execution.

8.3.2 Hardware Requirements

Although the Navigation Engineering Package set was designed to be as generic as possible, three hardware requirements limit its use.

1. Only applies to differential drive and complete wheeled robots. It assumes that moving the base is controlled by sending the desired speed command in the form of x speed, y speed, theta speed.
2. It requires a planar laser mounted on a mobile base for map building and positioning.
3. The Navigation Engineering Package set was developed on a square robot, so it performs best on robots that are close to square or round. While it can be used for robots of arbitrary shapes and sizes, large rectangular robots may have difficulty in tight spaces, such as doorways.

8.4   Basics of Using the Navigation Project Package Set

8.4.1   Installation and Configuration of the Navigation Project Package Set on the Robot

This section explains how to run the navigation project package set on any robot, mainly including: sending conversions using tf, publishing odometer information, publishing laser sensor data via ROS, and configuring the basic navigation project package set.

8.4.1.1   Construction of the Robot

The operation of the navigation project package set assumes that the robot is configured in a specific way. Fig. 8.3 shows an overview of this configuration. The white components are required components that have been implemented, the light gray components are optional components that have been implemented, and the dark gray components are components that must be created for each robot platform. The following provides an explanation of the prerequisites for using the Navigation Engineering package set and how each requirement is met.

1. ROS

The navigation project package set assumes that the robot uses ROS, the installation of which has been explained in the previous sections.

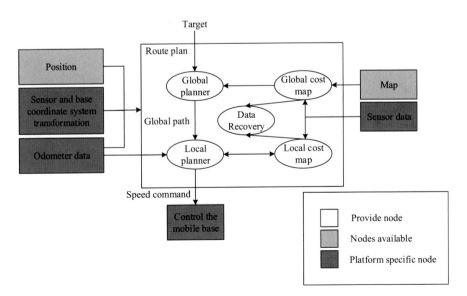

Fig. 8.3 Robot construction diagram

2. Conversion configuration

The navigation project package set requires the robot to use tf to publish information about the relationships between coordinate systems. Details of the configuration process can be found in Sect. 8.4.2.

3. Sensor information

The navigation project package set uses information from sensors to avoid obstacles, and it assumes that these sensors publish sensor_msgs/LaserScan or sensor_msgs/PointCloud messages through ROS. See Sect. 8.4.5 for information on publishing these messages through ROS. In addition, some sensors that already have a driver on the ROS can also be used for this purpose.

4. Odometer information

The navigation project package set requires that odometer information be posted using tf and nav_msgs/Odometry messages. You can refer to Sect. 8.4.4 of this book for information on how to publish odometer messages.

5. Base controller

The navigation project package set assumes that it is possible to send geometry_msgs/Twist messages based on the robot's base coordinate system to issue velocity commands via the "cmd_vel" topic, which means that there must be a node subscribed to the "cmd_vel " topic that can perform a velocity transformation of (vx, vy, vtheta) <==> (cmd_vel.linear.x, cmd_vel.linear.y, cmd_vel.angular.z) and convert it to a motor command to send to the mobile base.

6. Maps

The configuration of the navigation project package set does not require a map, but given the practicalities of the situation, it is still important to have a map for navigation. For more information on how to create a map, see the introduction to creating a map using Turtlebot in this book, or see http://wiki.ros.org/slam_gmapping/Tutorials/MappingFromLoggedData to learn how to create a 2D map.

8.4.1.2   Configure the Navigation Project Package

This section describes how to build and configure the navigation project package set on the robot. It is assumed that all of the above requirements for robot configuration are met. Specifically, this includes that the robot must publish information about the coordinate system using tf, receive sensor_msgs/LaserScan or sensor_msgs/PointCloud messages from all sensors used in the navigation package, and publish messages using tf and nav_msgs/Odometry, as well as receive and send velocity commands to the robot Base.

1. Create a project package

First create a project package in which we will store all the configuration and startup files for the navigation package. This project contains the dependencies for any project packages needed to satisfy the "Building the robot" section above, and contains the move_base package for the high-level interface to the navigation package. Next, choose a location for this package (robook/src or other) and run the following command.

```
$ cd robook_ws/src
$ catkin_create_pkg my_robot_name_2dnav move_base
my_tf_configuration_dep my_odom_configuration_dep
my_sensor_configuration_dep
```

This command will create a project package with the dependencies needed to run the navigation package on the robot. Here, both the navigation package and the dependencies need to be named on their own merits, and an example with an explanatory name is just given in this command.

2. Creating a robot startup configuration file

Now that we have created a workspace for all the configuration and launch files, we will next create a roslaunch file that will start the hardware needed to launch the robot and publish the tf needed for the robot. open the editor and paste the following code snippet into a file called my_robot_configuration.launch. You can replace the text "my_robot" with the name of the actual robot, but you must make a similar change to the launch file name. (The filename here is just an example, and since everyone's name may be different, the filename will be my_robot_configuration.launch throughout the rest of the text.)

```
<launch>

    <node pkg="sensor_node_pkg" type="sensor_node_type" name =
    "sensor_node_name" output="screen">
      <param name="sensor_param" value="param_value" />
    </node>
    <node pkg="odom_node_pkg" type="odom_node_type" name="odom_node"
    output="screen">
      <param name="odom_param" value="param_value" />
    </node>
    <node pkg = "transform_configuration_pkg" type =
    "transform_configuration_type" name =
    "transform_configuration_name" output = "screen">
        <param name="transform_configuration_param"
        value="param_value" />
    </node>

</launch>
```

We now have a template for the launch startup file, but still need to refine it for a specific bot. We will describe the changes that need to be made below.

```
<launch>
  <node pkg = "sensor_node_pkg" type = "sensor_node_type" name =
  "sensor_node_name" output = "screen">
```

Here we mention the sensors used by the robot for navigation, replacing sensor_node_pkg with the name of the ROS driver project package for the sensor, sensor_node_type with the type of the sensor driver, sensor_node_name with the name of the sensor node, and sensor_param with any parameter that the node can accept.

> Note: If you intend to use multiple sensors to send information to the navigation project package set, then they should be started together here.

```
  </node>
  <node pkg="odom_node_pkg" type="odom_node_type" name="odom_node"
  output="screen">
    <param name="odom_param" value="param_value" />
  </node>
```

Here, we start the base odometer. Again, the project package, type, name and parameter specification needs to be replaced with the parameters associated with the actual node being started.

```
    <param name="transform_configuration_param"
    value="param_value" />
  </node>
```

Here, we start the tf transform of the robot. Again, the project package, type, name and parameter specification needs to be replaced with the parameters associated with the actual node that is launched.

3. Configure costmap local_costmap and global_costmap

    The navigation package uses two kinds of cost maps to store information about obstacles in the physical world: one for global planning, i.e. creating long-term plans across the environment, and another for local planning and obstacle avoidance. There are some configuration options that we want both cost maps to follow, and some configuration options that we want to set on each map separately. Therefore, each of these three sections is described below: general configuration items, global configuration items, and local configuration items.

Note: What follows is just the basic configuration items for the costmap. For a complete configuration, please consult the costmap_2d documentation: http://wiki.ros.org/costmap_2d.
General configuration

Navigation packages use cost maps to store information about obstacles in the physical world. In order to do this correctly, we need to point the cost maps to the sensor topics they should listen to for updates. Create a file called costmap_common_params.yaml and fill it with the following:

```
obstacle_range: 2.5
raytrace_range: 3.0
footprint: [[x0, y0], [x1, y1], ... [xn, yn]]
#robot_radius: ir_of_robot
inflation_radius: 0.55

observation_sources: laser_scan_sensor point_cloud_sensor

laser_scan_sensor: {sensor_frame: frame_name, data_type: LaserScan,
topic: topic_name, marking: true, clearing: true}

point_cloud_sensor: {sensor_frame: frame_name, data_type:
PointCloud, topic: topic_name, marking: true, clearing: true}
```

Now, let's break down the code above.

```
obstacle_range: 2.5
raytrace_range: 3.0
```

These two parameters set the threshold for obstacle information to be put into the cost map. The obstacle_range parameter determines the maximum range of sensor readings that an obstacle can be put into the cost map. Here, we set it to 2.5m, which means that the robot will only update obstacle information centered on its base and within a 2.5m radius. raytrace_range parameter determines the extent of the blank area that the raytrace will reach. We set it to 3.0m, meaning the robot will attempt to clear the space 3.0m away in front of it based on the sensor readings.

```
footprint: [[x0, y0], [x1, y1], ... [xn, yn]]
#robot_radius: ir_of_robot
inflation_radius: 0.55
```

Here we set the occupancy area of the robot or the radius of the robot (if it is circular). In the case of specifying the occupancy area, it is assumed that the center of the robot is located at (0.0, 0.0) and that both clockwise and counterclockwise norms are supported. We will also set the expansion radius for the cost map. The swell radius should be set to the maximum distance from the obstacle, and the cost should be calculated from that distance. For example, setting the swell radius to 0.55m means that the robot will treat all paths that are 0.55m or further from an obstacle as having the same obstacle cost.

```
observation_sources: laser_scan_sensor point_cloud_sensor
```

The observation_sources parameter defines a list of sensors that pass information to the cost map. Each sensor is defined as shown below:

```
laser_scan_sensor: {sensor_frame: frame_name, data_type: LaserScan,
topic: topic_name, marking: true, clearing: true}
```

This line sets the parameters of the sensor mentioned in observation_sources, where laser_scan_sensor is defined. frame_name parameter should be set to the name of the sensor's coordinate system, and data_type parameter should be set to LaserScan or PointCloud, depending on which topic uses message. topic_name should be set to the name of the topic for which the sensor publishes data. The marking and clearing parameters determine whether the sensor adds obstacle information to the cost map, clears obstacle information from the cost map, or both.

Global cost map configuration

Below we will create a file which stores the configuration options for a specific global cost map. Open the file global_costmap_params.yaml with an editor and paste the following.

global_costmap:

```
global_frame: /map
robot_base_frame: base_link
update_frequency: 5.0
static_map: true
```

The global_frame parameter defines the coordinate system in which the cost map should be run, and in this case the /map coordinate system. The parameter robot_base_frame defines the robot base coordinate system that the cost map should refer to. update_frequency parameter determines the frequency (Hz) at which the cost map will run its update loop. static_map parameter determines whether the cost map should be initialized based on the map provided by the map_server cost map. If no existing map or map server is used, the static_map parameter needs to be set to false.

Local cost map configuration

Below we will create a file which stores the configuration options for a specific local cost map. Open the file local_costmap_params.yaml with an editor and paste the following.

local_costmap:

```
global_frame: odom
robot_base_frame: base_link
update_frequency: 5.0
publish_frequency: 2.0
static_map: false
rolling_window: true
width: 6.0
height: 6.0
resolution: 0.05
```

The global_frame, robot_base_frame, update_frequency, and static_map parameters are the same as described in the global cost map configuration section. The publish_frequency parameter determines the rate (in Hz) at which the cost map publishes visualizations. Setting the rolling_window parameter to true means that the cost map will always be centered on the robot when it is moving. width, height, and resolution parameters set the width (in meters), height (in meters), and resolution (in meters/cell) of the cost map. Note that the resolution of this grid can be different from the resolution of the static map, but in most cases we prefer to set them to the same resolution.

4. Configuration of the basic local planner

The base_local_planner is responsible for calculating the commands for the speed, sending these commands to the robot's mobile base, giving an advanced plan. We need to set some configuration options based on the robot's specifications. Open a file called base_local_planner_params.yaml and paste the following.

Note: This section covers only the basic configuration options for TrajectoryPlanner. Please refer to the base_local_planner documentation for the full set of configuration options: http://wiki.ros.org/base_local_planner.

```
TrajectoryPlannerROS:
  max_vel_x: 0.45
  min_vel_x: 0.1
  max_vel_theta: 1.0
  min_in_place_vel_theta: 0.4

  acc_lim_theta: 3.2
  acc_lim_x: 2.5
  acc_lim_y: 2.5

  holonomic_robot: true
```

The first part of the above parameters defines the velocity limit of the robot. The second part defines the acceleration limit of the robot.

5. Create a Launch startup file for the navigation package

Now that we have all the configuration files ready to go, we next need to merge everything into the launch file of the navigation package. Open the file move_base. launch and edit it to.

```
<launch>

  <master auto="start"/>
  <! -- Run the map server -->
  <node name="map_server" pkg="map_server" type="map_server"
args="$(find my_map_package)/my_map.pgm my_map_resolution"/>

  <! --- Run AMCL -->
  <include file="$(find amcl)/examples/amcl_omni.launch" />

  <node pkg="move_base" type="move_base" respawn="false"
name="move_base" output="screen">
    <rosparam file="$(find my_robot_name_2dnav)/ costmap_common_params.
yaml" command="load" ns="global_costmap" />
    <rosparam file="$(find my_robot_name_2dnav)/ costmap_common_params.
yaml" command="load" ns="local_costmap" />
     <rosparam file="$(find my_robot_name_2dnav)/ local_costmap_params.
yaml" command="load" />
    <rosparam file="$(find my_robot_name_2dnav)/ global_costmap_params.
yaml" command="load" />
    <rosparam file="$(find my_robot_name_2dnav)/
base_local_planner_params.yaml" command="load" />
  </node>

</launch>
```

In that file, you need to change the map server to point to an existing map, and change some of the config file paths to your own saved paths. If you have a differentially driven robot, amcl_omni.launch should be changed to amcl_diff. launch. For how to create a map, see: http://wiki.ros.org/slam_gmapping/Tutorials/ MappingFromLoggedData, or use the Turtlebot bot platform (refer to Sect. 8.5.1).

6. AMCL configuration (amcl)

AMCL has many configuration options that affect the performance of the positioning. For more information on AMCL please refer to http://wiki.ros.org/amcl.

8.4.1.3 Run the Navigation Package

Now that everything is set up, we can run the navigation package. To do this, the robot should come with two terminals on it. In one terminal, start the my_robot_configuration.launch file; in the other, start the move_base.launch file you just created.

```
Terminal 1
$ roslaunch my_robot_configuration.launch
Terminal 2
$ roslaunch move_base.launch
```

At this point, if no errors are reported, then the navigation package is running.

8.4.2 Robot tf Configuration

8.4.2.1 The Concept of tf Conversion

Many ROS packages require the use of the tf software library to publish the robot's transformation tree. At the abstraction level, the transformation tree defines offsets based on translations and rotations between different coordinate systems. To illustrate this more graphically, consider the example of a simple robot. This robot has a mobile base on which a laser is mounted. For this robot, we define two coordinate systems: one corresponding to the center point of the robot base and the other corresponding to the center point of the laser mounted on top of the base. For ease of reference, we refer to the coordinate system on the mobile base as base_link (it is important for navigation to place it at the center of rotation of the robot) and refer to the coordinate system attached to the laser as base_laser.

Suppose we have some data from the laser in the form of distance from the laser's center point. In other words, we have some data in the base_laser coordinate system. Now suppose we want to use this data to help move the base to avoid obstacles in the physical world. To do this, we need a way to convert the laser scan data we receive from the base_laser coordinate system to the base_link coordinate system. Essentially, we need to define the relationship between the base_laser and base_link coordinate systems.

In defining this relationship, assume we know that the laser is mounted 10 cm in front of and 20 cm above the center point of the moving base, as shown in Fig. 8.4. This gives us a translation offset that relates the base_link and base_laser coordinate systems. Specifically, to get data from the base_link coordinate system to the base_laser coordinate system, we must use the transformation (x: 0.1m, y: 0.0m, z: 0.2m); to get data from the base_laser coordinate system to the base_link coordinate system, we must apply the opposite transformation (x: −0.1 m, y: 0.0 m, z: −0.20 m).

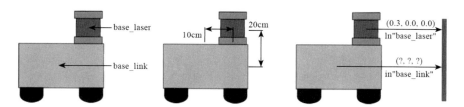

Fig. 8.4 Schematic representation of the relationship between the base_laser and base_link coordinate systems

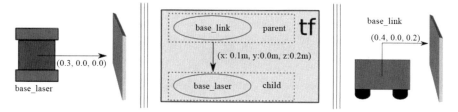

Fig. 8.5 Example of a tf conversion tree

We could choose to manage this relationship ourselves, which would mean storing and applying conversions between coordinate systems when necessary, but this becomes more and more work as the number of coordinate systems increases. Fortunately, we don't have to do this work ourselves; tf will define the relationship between base_link and base_laser and manage the conversion between the two coordinate systems.

In order to use the relationships between the base_link and base_laser coordinate systems defined and stored by tf, we need to add them to the transformation tree. Conceptually, each node in the transformation tree corresponds to a coordinate system, and each edge corresponds to the transformation that needs to be applied when moving from the current node to its children. tf uses a tree structure and assumes that all edges in the tree point from the parent node to the children to ensure that any two coordinate systems can be joined together in a single traversal.

A simple example of creating a transformation tree is given next. We will create two nodes, one for the base_link coordinate system and the other for the base_laser coordinate system, as shown in Fig. 8.5. To create an edge between them, we first need to decide which node is the parent and which node is the child. Note that this distinction is important because tf assumes that all transformations are from parent to child nodes. We choose the base_link coordinate system as the parent because it makes the most sense to relate to the base_laser coordinate system via the base_link coordinate system when adding other parts/sensors to the robot. This means that the transformation corresponding to the edge connecting base_link to base_laser is (x: 0.1 m, y: 0.0 m, z: 0.2 m). Once the transformation tree is set up, the conversion of the laser scan data received in the base_laser coordinate system to the base_link coordinate system can be achieved by calling the tf library. The robot can use this information to derive the laser scan data in the base_link coordinate system and safely perform path planning and avoid obstacles in the surrounding environment.

8.4.2.2 Writing the Code

As mentioned above, suppose we want to take points in the base_laser coordinate system and convert them to the base_link coordinate system. The first thing we need to do is create a node that will be responsible for posting the conversion on the system. Next, a node must be created to listen to the conversion data posted on ROS and use it to convert a point. We first create a package, put the source code in it, and give it a simple name like ch8_setup_tf. This package depends on roscpp, tf, and geometry-msg.

```
$ cd robook_ws/src
$ catkin_create_pkg ch8_setup_tf roscpp tf geometry_msgs
```

8.4.2.3 Dissemination of a Conversion

Now we have our package, the next step is to create a node to do the job of propagating the base_laser to base_link conversion via ROS. In the ch8_setup_tf package you just created, start the editor you wish to use and paste the following code into the tf_broadcaster.cpp file in the src/ folder.

```
1 #include <ros/ros.h>
2 #include <tf/transform_broadcaster.h>
3
4 int main(int argc, char** argv){
5     ros::init(argc, argv, "robot_tf_publisher");
6     ros::NodeHandle n;
7
8     ros::Rate r(100);
9
10    tf::TransformBroadcaster broadcaster;
11
12    while(n.ok()){
13        broadcaster.sendTransform(
14        tf::StampedTransform(
15        tf::Transform(tf::Quaternion(0, 0, 0, 1), tf::Vector3(0.1,
           0.0, 0.2)),
16        ros::Time::now(), "base_link", "base_laser"));
17        r.sleep();
18    }
19 }
```

Now, we illustrate the code for the above base_link to base_laser conversion.

Line 2: The tf package provides an implementation of tf::TransformBroadcaster to simplify the task of publishing transformations. To use TransformBroadcaster, the tf/transform_broadcaster.h header file needs to be included.

Line 10: We create a TransformBroadcaster object, which will be used later to send the base_link to base_laser conversion.

Lines 13 to 16: This part is the real work. Sending the transformation with TransformBroadcaster takes five parameters. The first is the btQuaternion parameter, which indicates any rotational transformations that occur between the two coordinate systems. In this case, we send a quaternion consisting of pitch, roll, and yaw values, among others. The second is the btVector3 parameter for any transformation we want to apply. However, in this case we want to apply a translation, so a btVector3 is created, corresponding to the case where base_laser is 10 cm x offset and 20 cm z offset from the robot base_link. The third is the ros::Time::now() parameter, the timestamp is used to mark the transformation being posted. The fourth is the base_link parameter, which indicates the name of the parent node of the connection created. The fifth is the base_laser parameter, which indicates the name of the child node of the connection created.

8.4.2.4 Use a Conversion

We have created a node that is able to publish the base_laser to base_link conversion via ROS. Now, we will write a node that uses that transformation to get a point in the base_laser coordinate system and convert it to a point in the base_link coordinate system. Similarly, in the src/ folder in the ch8_setup_tf package, create a file called tf_listener.cpp and paste the following:

```
1 #include <ros/ros.h>
2 #include <geometry_msgs/PointStamped.h>
3 #include <tf/transform_listener.h>
4
5 void transformPoint(const tf::TransformListener& listener){
6     //we'll create a point in the base_laser frame that we'd like to
      transform to the base_link frame
7     geometry_msgs::PointStamped laser_point;
8     laser_point.header.frame_id = "base_laser";
9
10    //we'll just use the most recent transform available for our simple
      example
11    laser_point.header.stamp = ros::Time();
12
13    //just an arbitrary point in space
14    laser_point.point.x = 1.0;
15    laser_point.point.y = 0.2;
16    laser_point.point.z = 0.0;
17
18    try{
19        geometry_msgs::PointStamped base_point;
```

```
20        listener.transformPoint("base_link", laser_point,
          base_point);
21
22        ROS_INFO("base_laser: (%.2f, %.2f, %.2f) ----> base_link:
          (%.2f, %.2f, %.2f) at time %.2f",
23    laser_point.point.x, laser_point.point.y, laser_point.point.z,
24    base_point.point.x, base_point.point.y, base_point.point.z,
          base_point.header.stamp.toSec());
25    }
26    catch(tf::TransformException& ex){
27        ROS_ERROR("Received an exception trying to transform a point
          from \"base_laser\" to \"base_link\": %s", ex.what());
28    }
29 }
30
31 int main(int argc, char** argv){
32    ros::init(argc, argv, "robot_tf_listener");
33    ros::NodeHandle n;
34
35    tf::TransformListener listener(ros::Duration(10));
36
37    //we'll transform a point once every second
38    ros::Timer timer = n.createTimer(ros::Duration(1.0), boost::
          bind(&transformPoint, boost::ref(listener)));
39
40    ros::spin();
41
42 }
```

Next, we analyze the key points in the above code.

Line 3: The tf/transform_listener.h header file is included here, as it is required to create tf::TransformListener. The TransformListener object automatically subscribes to the topic of transformation messages via ROS and manages all incoming transformation data.

Line 5: Create a function, given a TransformListener, get a point in the base_laser coordinate system and transform it to the base_link coordinate system. This function will be called back in the main() of the program by ros::Timer, triggered once per second.

Lines 6–16: Create a point named geometry_msgs::PointStamped. The stamped at the end of the message name indicates that it contains a header that allows the timestamp and coordinate system ID (frame_id) to be associated with the message. We set the stamped field of the laser_point message to ros::Time(), which is a special time value that allows us to request the latest available transformation from the TransformListener. For the frame_id field of the header, we set it to base_laser, since the point we created is in the base_laser coordinate system. Finally, we set some data for the point, picking values of x: 1.0, y: 0.2, z: 0.0.

Lines 18–25: Now we have a point in the base_laser coordinate system, and we want to convert it to the base_link coordinate system. To do this, we use the TransformListener object and call transformPoint() with three parameters: the name of the coordinate system of the point to be transformed (in this case, it is base_link), the point to be transformed, and the location where the point will be stored after the transformation. Thus, after calling transformPoint(), base_point and laser_point hold the same information, except that base_point is now in the base_link coordinate system.

Lines 26–28: If for some reason the conversion from base_laser to base_link is not available (perhaps tf_broadcaster is not running), the TransformListener may raise an exception when TransformPoint() is called. To ensure proper handling, the exception is caught and an error is printed out for the user.

8.4.2.5   Compile the Code

After writing the example code, we need to compile the code. Before compiling, you need to open the CMakeLists.txt file that was automatically generated when you created the project package, and add the following line to the bottom of the file:

```
add_executable(tf_broadcaster src/tf_broadcaster.cpp)
add_executable(tf_listener src/tf_listener.cpp)
target_link_libraries(tf_broadcaster ${catkin_LIBRARIES})
target_link_libraries(tf_listener ${catkin_LIBRARIES})
```

Next, save the file and compile the project package:

```
$ cd robook_ws
$ catkin_make
```

8.4.2.6   Run the Code

After compiling the code, we try to run the code. In this part, three terminals need to be opened.

```
The first terminal
$ roscore
Second terminal
Run our tf_broadcaster:
$ rosrun ch8_setup_tf tf_broadcaster
Third terminal
Run tf_listener to convert the simulation points from the base_laser
coordinate system to the base_link coordinate system.
$ rosrun ch8_setup_tf tf_listener
```

If all is well, you will see a point per second converted from the base_laser coordinate system to the base_link coordinate system.

The next step is to replace the PointStamped used in this example with a sensor stream from ROS, which is explained in Sect. 8.4.5 of this book.

8.4.3   Basic Navigation Debugging Guide

8.4.3.1   Robot Navigation Preparation

When debugging the navigation stack on a new robot, most of the problems we encounter are located in areas outside of the local planner debug parameters. The robot's odometer, localization, sensors, and other prerequisites for running navigation effectively are often problematic. So, the first thing to do is to make sure the robot itself is ready for navigation, which includes a check of three components: distance sensors, odometer, and localization.

1. Distance sensor
   If the robot cannot get information from a distance sensor (such as a laser), then the navigation system will not work. Therefore, one needs to make sure that the sensor information can be viewed in rviz. When the information looks correct, it enters the system at the expected rate.

2. Odometer
   Often, we have difficulty getting an accurate position of the robot. Its position information is constantly lost, and we spend a lot of time working with AMCL parameters, only to find that the real culprit is the robot's odometer. Therefore, we usually need to run two complete and thorough checks to ensure that the robot's odometer is working reliably.

   The first check is to see if the rotation of the odometer makes sense. Turn on rviz for the first time, set the frame to odom, display the laser scan function provided by the robot, set the decay time of the subject to high (about 20 seconds), and perform a rotation in place. Then, observe how closely the scan matches in subsequent rotations. Ideally, the scans will overlap exactly, but there will be some rotational drift, so one needs to ensure that the scan deviation values are between 1 and 2.

   Another check is to perform a full inspection of the odometer by placing the robot a few meters away from the wall. Set up the rviz in the manner described above, then drive the robot towards the wall and observe the thickness of the wall as scanned by the aggregated laser in the rviz. Ideally, the scan of the wall only needs to ensure that it is no more than a few centimeters thick. If the robot is driven one meter towards the wall, but the scan spreads to more than half a meter, then there may be a problem with this odometer.

3. Positioning
   Assuming that the odometer and laser scanner are working properly, mapping and debugging the AMCL is usually not too difficult. First, run gmapping or karto and manipulate the robot to generate a map. Then, use the map in AMCL and make sure the robot stays positioned. If the odometer of the running robot is not working well, you can make changes to the odometer model parameters of AMCL. A good test of the whole system is to make sure that the laser scan and the map are visible in the map of rviz and the laser scan matches well with the

environment map. A reasonable initial pose for the robot should be set in rviz before the robot makes the desired movement.

8.4.3.2 Cost Map

If it is determined that the robot meets the prerequisites for navigation, then the next step is to ensure that the cost map is set up and configured correctly. The following are some recommendations when debugging the cost map.

Make sure to set the expected_update_rate parameter for each watch source based on the actual rate posted by the sensor. A larger tolerance is given here, setting the check period to twice the expected value; the better case is to receive a warning from the navigation when the sensor's rate is much lower than expected.

Set the transform_tolerance parameter appropriately for the system. Use tf to check the expected latency of the transformation from the base_link coordinate system to the map coordinate system. It is common to use tf_monitor to see the latency of the system, usually with the parameter set to off. Alternatively, if tf_monitor reports a large latency, you can see what is causing the latency. Sometimes we find it is a matter of how to post the transformation for a given robot.

For robots lacking processing power, one could consider turning off the map_update_rate parameter. However, this needs to be considered in light of the fact that doing so will cause a delay in the rate at which sensor data enters the cost map, which in turn slows down the robot's reaction to obstacles.

The publish_frequency parameter is useful for implementing costly map visualizations in rviz. However, for large global maps, this parameter can cause slow operation. In a production system, one could consider reducing the speed of cost map publishing and set this very low when very large maps need to be visualized.

Whether to use the voxel_grid or costmap model for cost maps depends heavily on the robot's sensor suite. Debugging a 3D-based costmap can be more complicated because of the unknown space that needs to be considered. If the robot used has only one planar laser, the costmap model can be used for mapping.

Sometimes it is useful to be able to run the navigation separately in the odometer coordinate system. The easiest way to do this is to copy the local_costmap_params. yaml file to overwrite the global_costmap_params.yaml file, and change the width and height of the map to about 10m. This is a very simple way to optimize navigation if you want to not rely on positioning performance.

The resolution of the map is chosen based on the size and processing power of the robot. For a robot that has a lot of processing power and needs to fit into a small space (like PR2), it could use a fine-grained map and set the resolution to 0.025 m). For a robot like roomba, it could set the resolution to 0.1 m to reduce the computational load.

The rviz is a good way to verify that the cost map is working properly. It is common to look at obstacle data from the cost map and make sure that the obstacle data is consistent with the map and laser scan when driving the robot under joystick control. This is a complete check to ensure that the sensor data is entering the cost map in a reasonable manner. If tracking in unknown space with a robot, most often the cost map is drawn using the voxel_grid model, which visualizes the unknown space and ensures that it is cleared in a reasonable manner. To see if the obstacles are cleared correctly from the cost map, walk in front of the robot and see if it can successfully see you and clear from the map. For more information on costmap publishing to rviz, check out the navigation section of the rviz tutorial.

When the navigation project package set is only running cost maps, it is best to check the system load. This means opening the move_base node, but not sending the target and checking the load. If the computer gets bogged down at this point, then to run the planner, some CPU-saving parameter tuning is required.

8.4.3.3 Local Planner

If one is satisfied with the results obtained through the cost map, then one can next optimize the parameters in the local planner. For robots with reasonable acceleration limits, dwa_local_planner can be used; for robots with lower acceleration limits, it is better to consider the acceleration limits at each step and use base_local_planner. Debugging dwa_local_planner is more convenient than debugging base_local_planner because its parameters are dynamically configurable. The following are some recommendations for the planner.

For both planners, dwa_local_planner and base_local_planner, the most important thing is to set the acceleration limit parameters correctly for the given robot. If these parameters are off, one can only expect the robot to behave sub-optimally. If it is not known what the acceleration limits are for the robot, you can write a script to make the motor run at maximum translational and rotational speed for a period of time, and then look at the odometer feedback for the speed (assuming the odometer gives a reasonable estimate) to get the acceleration limits. Setting these parameters appropriately can save a lot of time.

If the robot has a low acceleration limit, you need to ensure that you run base_local_planner with dwa set to false. After setting dwa to true, you need to update the value of the vx_samples parameter to between 8 and 15, depending on the effective processing power. This will generate non-circular curves during the presentation.

If the robot used has poor positioning capabilities, the target tolerance parameter can be set higher than it would otherwise be. If the robot has a high minimum rotation speed, the rotation tolerance can also be increased to avoid oscillations at the target point.

If a low resolution is used because of the CPU, the sim_granularity parameter can be increased to save some cycles.

The path_distance_bias and goal_distance_bias parameters on the planner (for base_local_planner, these are called pdist_scale and gdist_scale) are generally rarely changed. When they are changed, it is usually because of a desire to limit the freedom of the local planner to leave the planned path and work with the global planner rather than with NavFn. Turning the path_distance_bias parameter up will bring the robot closer to the path, but at the cost of moving quickly towards the target. If this weight is set too high, the robot will refuse to move because the cost of moving is higher than the cost of staying at some location on the path.

If you want to reason about the cost function in an intelligent way, you need to set the meter_scoring parameter to true. At this time, the distance in the cost function is measured in meters rather than in cells. This also means that the cost function can be tuned for one map resolution and expect reasonable behavior when moving to other locations as well. In addition, setting the publish_cost_grid parameter to true allows visualization of the cost function generated by the local planner in rviz. Given the cost function in meters, a trade-off can be made between the cost of moving 1m towards the target and the distance from the planned path. This helps to better understand how to perform debugging.

The trajectory is calculated from the end point. Setting the sim_time parameters to different values can have a big impact on the behavior of the robot. You can generally set this parameter to 1 to 2 s. If you set the parameter to a higher value, the trajectory will be smoother, but make sure that the minimum velocity multiplied by sim_period is less than twice the tolerance value to the target; otherwise, the robot will rotate at a position other than the target position instead of moving towards the target.

Accurate trajectory simulation also relies on obtaining a reasonable velocity estimate from the odometer. This is because both dwa_local_planner and base_local_planner use this velocity estimate, along with the robot's acceleration limits, to determine the feasible velocity space for a planning cycle. While the speed estimate obtained from the odometer is not necessarily perfect, it is important to ensure that it is at least close to optimal.

8.4.4 Release of Odometer Measurements Via ROS

The navigation engineering package uses tf to determine the robot's position in the global coordinate system and to correlate sensor data with a static map. However, tf does not provide any information about the robot's velocity. Therefore, the navigation engineering package requires all odometers to issue a message via ROS that translates and contains velocity information for nav_msgs/Odometry. This section provides an example of publishing an odometer message for a navigation engineering package set. It consists of publishing a nav_msgs/Odometry message via ROS and converting from the odom coordinate system to the base_link coordinate system via TF.

8.4.4.1 nav_msgs/Odometry Messages

The nav_msgs/Odometry message stores an estimate of the robot's position and velocity in free space.

```
# This represents an estimate of a position and velocity in free space.
# The pose in this message should be specified in the coordinate frame
given by header.frame_id.
# The twist in this message should be specified in the coordinate frame
given by the child_frame_id.
Header header
string child_frame_id
geometry_msgs/PoseWithCovariance pose
geometry_msgs/TwistWithCovariance twist
```

The pose in this message corresponds to the estimated position of the robot in the odometer coordinate system, with an optional covariance for determining that pose estimate. The twist in this message corresponds to the robot's velocity in a subcoordinate system (typically a coordinate system based on a moving base) with an optional covariance for determining that velocity estimate.

8.4.4.2   Use tf to Post Odometer Conversions

As discussed in Section 8.4.2, the tf software library is responsible for managing the relationships between the coordinate systems associated with the robot in the transformation tree. Therefore, any odometer must publish information about the coordinate systems it manages.

8.4.4.3   Writing the Code

The following code is used to issue nav_msgs/Odometry messages via ROS, and to perform a coordinate transformation of a simulated robot traveling in a circle by using tf.

First, add the dependencies to the package.xml file in the project package.

```
<build_depend>tf</build_depend>
<build_depend>nav_msgs</build_depend>
<build_export_depend>tf</build_export_depend>
<exec_depend>tf</exec_depend>
<exec_depend>nav_msgs</exec_depend>
```

In the src/ folder of the ch8_setup_tf package, create a file called odometry_publisher.cpp and paste the following.

```
1 #include <ros/ros.h>
2 #include <tf/transform_broadcaster.h>
3 #include <nav_msgs/Odometry.h>
4
5 int main(int argc, char** argv){
6     ros::init(argc, argv, "odometry_publisher");
7
```

```
8    ros::NodeHandle n;
9    ros::Publisher odom_pub = n.advertise<nav_msgs::Odometry>
     ("odom", 50);
10   tf::TransformBroadcaster odom_broadcaster;
11
12   double x = 0.0;
13   double y = 0.0;
14   double th = 0.0;
15
16   double vx = 0.1;
17   double vy = -0.1;
18   double vth = 0.1;
19
20   ros::Time current_time, last_time;
21   current_time = ros::Time::now();
22   last_time = ros::Time::now();
23
24   ros::Rate r(1.0);
25   while(n.ok()){
26
27       ros::spinOnce(); // check for incoming messages
28       current_time = ros::Time::now();
29
30       //compute odometry in a typical way given the velocities of the
         robot
31       double dt = (current_time - last_time).toSec();
32       double delta_x = (vx * cos(th) - vy * sin(th)) * dt;
33       double delta_y = (vx * sin(th) + vy * cos(th)) * dt;
34       double delta_th = vth * dt;
35
36       x += delta_x;
37       y += delta_y;
38       th += delta_th;
39
40       //since all odometry is 6DOF we'll need a quaternion created
         from yaw
41       geometry_msgs::Quaternion odom_quat = tf::
         createQuaternionMsgFromYaw(th);
42
43       //first, we'll publish the transform over tf
44       geometry_msgs::TransformStamped odom_trans;
45       odom_trans.header.stamp = current_time;
46       odom_trans.header.frame_id = "odom";
47       odom_trans.child_frame_id = "base_link";
48
49       odom_trans.transformation.translation.x = x;
50       odom_trans.transformation.translation.y = y;
51       odom_trans.transform.translation.z = 0.0;
52       odom_trans.transform.rotation = odom_quat;
53
54       //send the transform
55       odom_broadcaster.sendTransform(odom_trans);
56
```

```
57          //next, we'll publish the odometry message over ROS
58          nav_msgs::Odometry odom;
59          odom.header.stamp = current_time;
60          odom.header.frame_id = "odom";
61
62          //set the position
63          odom.pose.pose.position.x = x;
64          odom.pose.pose.position.y = y;
65          odom.pose.pose.position.z = 0.0;
66          odom.pose.pose.orientation = odom_quat;
67
68          //set the velocity
69          odom.child_frame_id = "base_link";
70          odom.twist.twist.linear.x = vx;
71          odom.twist.twist.linear.y = vy;
72          odom.twist.twist.angular.z = vth;
73
74          //publish the message
75          odom_pub.publish(odom);
76
77          last_time = current_time;
78          r.sleep();
79      }
80 }
```

Next, we explain the important parts of this code in detail.

Lines 2–3: Since we're posting a conversion from the odom coordinate system to the base_link coordinate system and a nav_msgs/odometry message, we need to include the relevant header files.

Lines 9–10: We need to create ros::Publisher and tf::TransformBroadcaster in order to send messages via ROS and TF respectively.

Lines 12–14: We assume that the robot initially starts its action from the origin of the odom coordinate system.

Lines 16–18: Here we set some velocities so that the base_link coordinate system moves in the x-direction at the rate of 0.1m/s, in the y-direction at the rate of -0.1m/s, and in the z-direction at the rate of 0.1rad/s under the odom coordinate system, which will cause the robot to travel in a circle.

Line 24: In this example, we post the odometer information at 1Hz for inspection purposes. In practice, most systems will want to post odometer information at a higher rate.

Lines 30–38: Here we will update the odometer information based on the set constant speed. Of course, a real odometer system would integrate the calculated speed.

Lines 40–41: We usually try to use a 3D version of the message in the system to allow the 2D and 3D components to work together when appropriate and to keep the number of messages created as small as possible. Therefore, it is necessary to convert the odometer yaw values to quaternions and send them out. Fortunately, tf provides features that allow easy creation of quaternions from yaw values and easy access to yaw values from quaternions.

Lines 43–47: Here we create a conversion message to be sent via tf. We want to post a transformation from the odom coordinate system to the base_link coordinate system at current_time, so the message's header and child_frame_id (child coordinate system ID) will be set, ensuring that odom is used as the parent coordinate system and base_link is used as the child coordinate system.

Lines 49–55: Here we populate the transformation message from the mileage data and then use TransformBroadcaster to send the transformation message.

Lines 57 to 60: We also need to publish the nav_msgs/Odometry message to navigate the project package set for velocity information. We will set the message headers to current_time and odom coordinate system.

Lines 62–72: Here the message is populated with odometer data and sent out. We set the child_frame_id of the message to the base_link coordinate system, since this is the coordinate system in which the speed information is sent.

8.4.4.4   Compile and Run

After writing the example code, we need to compile the code. Before compiling, you need to open the CMakeLists.txt file that was automatically generated when the project package was created and add the following line to the bottom of the file.

```
add_executable(odometry_publisher src/ odometry_publisher.cpp)
target_link_libraries(odometry_publisher ${catkin_LIBRARIES})
```

Next, save the file and compile the project package:

```
$ cd robook_ws
$ catkin_make
```

After compiling the code, we try to run the code. For this part, you need to open three terminals.

```
The first terminal
$ roscore
Second terminal
Run our odometry_publisher:
$ rosrun ch8_setup_tf odometry_publisher
Third terminal
Run rviz:
$ rosrun rviz rviz
```

Some settings are needed in rviz: change fixed frame to odom; click Add, find Odometry, and add it; change Topic under Odometry to /odom and see the result in Fig. 8.6 after completing the settings, with the simulated robot driving inside the circle and performing the coordinate transformation.

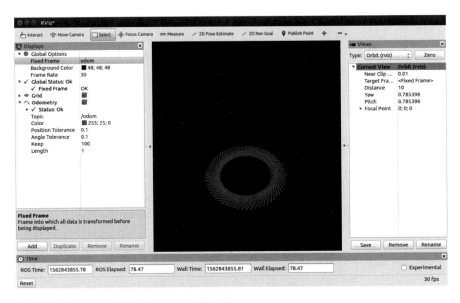

Fig. 8.6 odometry_publisher's run results in rviz

8.4.5 Publishing Sensor Data Streams Via ROS

Correctly posting data from sensors via the ROS is important for the safe operation of the navigation engineering package set. If the navigation engineering package set does not receive information from the robot's sensors, it will drive blindly and likely hit something. There are many sensors that can be used to provide information to the Navigation Engineering Package set, including lasers, cameras, sonar, infrared, collision sensors, etc. However, the current navigation engineering package set only accepts sensor data published using the sensor_msgs/LaserScan message type or the sensor_msgs/PointCloud message type.

This section provides examples of sending two types of sensor streams (i.e., sensor_msgs/LaserScan messages and sensor_msgs/PointCloud messages) via ROS.

8.4.5.1 ROS Message Headers

As with many messages sent via ROS, both the sensor_msgs/LaserScan and sensor_msgs/PointCloud message types contain tf coordinate system and time related information. To standardize how this information is sent, the Header message type is used for the fields in all such messages.

The three fields in the Header type are shown below.

```
#Standard metadata for higher-level flow data types
#sequence ID: consecutively increasing ID
```

```
uint32 seq
#Two-integer timestamp that is expressed as:
# * stamp.secs: seconds (stamp_secs) since epoch
# * stamp.nsecs: nanoseconds since stamp_secs
# time-handling sugar is provided by the client library
time stamp
#Frame this data is associated with
# 0: no frame
# 1: global frame
string frame_id
```

The seq field corresponds to an identifier whose value is automatically incremented as messages from a given publisher are sent. The stamp field stores information about the time associated with the data in the message. For example, in the case of laser scanning, stamp may correspond to the time at which the scan started. The frame_id field stores information about the tf coordinate system associated with the data in the message. In the case of laser scanning, this will be set to the laser data's coordinate system.

8.4.5.2   Publish LaserScan Messages on ROS

LaserScan Message

For robots that use laser scanners, ROS provides a special message type LaserScan in the sensor_msgs package to hold information for a given scan. As long as the data returned from the scanner can be formatted to fit the message, LaserScan messages make it easy for the code to handle any laser scan. Before discussing how these messages are generated and published, let's look at the message specification.

```
# Measured laser scan angle, counterclockwise is positive
# 0 degrees of the device coordinate system facing forward (along the X-
axis)
```

```
Header header
float32 angle_min # The start angle of the scan [in radians].
float32 angle_max # The end angle of the scan [in radians].
float32 angle_increment # The distance between the measured angles [in
radians]
float32 time_increment # The time between measurements [seconds]
float32 scan_time # Time between scans [seconds]
float32 range_min # Minimum measured distance [meters]
float32 range_max # Maximum distance to be measured [meters]
float32[] ranges # Measured distance data [meters] (Note: values less
than range_min or greater than range_max should be discarded)
float32[] intensities # Intensity data [device-specific units]
```

The names/comments above clearly show most of the fields in the message. To be more specific, let's write a simple laser data publisher in order to illustrate how they work.

Write Code to Publish a LaserScan Message

Publishing LaserScan messages via ROS is fairly simple. In the src/ folder of the ch8_setup_tf package, create a file called laser_scan_publisher.cpp and paste the following.

```
1 #include <ros/ros.h>
2 #include <sensor_msgs/LaserScan.h>
3
4 int main(int argc, char** argv) {
5     ros::init(argc, argv, "laser_scan_publisher");
6
7     ros::NodeHandle n;
8     ros::Publisher scan_pub = n.advertise<sensor_msgs::LaserScan>
      ("scan", 50);
9
10    unsigned int num_readings = 100;
11    double laser_frequency = 40;
12    double ranges[num_readings];
13    double intensities[num_readings];
14
15    int count = 0;
16    ros::Rate r(1.0);
17    while(n.ok()){
18        //generate some fake data for our laser scan
19        for(unsigned int i = 0; i < num_readings; ++i){
20            ranges[i] = count;
21            intensities[i] = 100 + count;
22        }
23        ros::Time scan_time = ros::Time::now();
24
25        //populate the LaserScan message
26        sensor_msgs::LaserScan scan;
27        scan.header.stamp = scan_time;
28        scan.header.frame_id = "laser_frame";
29        scan.angle_min = -1.57;
30        scan.angle_max = 1.57;
31        scan.angle_increment = 3.14 / num_readings;
32        scan.time_increment = (1 / laser_frequency) / (num_readings);
33        scan.range_min = 0.0;
34        scan.range_max = 100.0;
35
36        scan.ranges.resize(num_readings);
37        scan.intensities.resize(num_readings);
38        for(unsigned int i = 0; i < num_readings; ++i){
39            scan.ranges[i] = ranges[i];
```

```
40                    scan.intensities[i] = intensities[i];
41            }
42
43          scan_pub.publish(scan);
44          ++count;
45          r.sleep();
46      }
47 }
```

Now, let's analyze the important parts of the above code in detail.

Line 2: Contains the sensor_msgs/LaserScan message we want to send.

Line 8: Create a ros::Publisher for sending LaserScan messages via ROS.

Lines 10–13: Create storage variables for the virtual data that will be used to populate the scan, which an actual application will pull from their laser drive.

Lines 18–23: Fill the virtual laser data with values increasing by 1 per second.

Lines 25–41: Create a scan_msgs::LaserScan message and fill it with the generated ready-to-send data.

Line 43: Post this message on the ROS.

Run to Display LaserScans

After writing the sample code, we need to compile it. Before compiling, you need to open the CMakeLists.txt file that was automatically generated when you created the project package and add the following line to the bottom of the file.

```
add_executable(laser_scan_publisher src/ laser_scan_publisher.cpp)
target_link_libraries(laser_scan_publisher ${catkin_LIBRARIES})
```

Next, save the file and compile the project package:

```
$ cd robook_ws
$ catkin_make
```

After compiling the code, we try to run the code. In this part, three terminals need to be opened.

```
The first terminal
$ roscore
Second terminal
Run our odometry_publisher:
$ rosrun ch8_setup_tf laser_scan_publisher
Third terminal
Running rviz.
$ rosrun rviz rviz
```

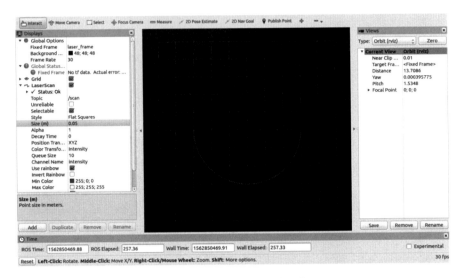

Fig. 8.7 Showing the results of laser_scan_publisher runs in rviz

Some settings are required in rviz: change fixed frame to laser_frame; click Add, find LaserScan and add it; change the Topic under LaserScan to /scan. Once the settings are completed you can see the result shown in Fig. 8.7, which is the published virtual laser data.

Publish Point Clouds on ROS PointClouds

1. Point clouds message

To store and share data for a series of points on the space, ROS provides sensor_msgs/PointCloud messages, as follows.

#This message contains a set of 3d points, and optional additional information about each point

#Each Point32 should be interpreted as a 3d point in the frame given in the header
Header header
geometry_msgs/Point32[] points # 3d point array
ChannelFloat32[] channels # The number of elements in each channel should be the same as the array of points, and the data in each channel should correspond to each point 1:1

This message is used to support point arrays in 3D and any associated data stored as channels. For example, PointCloud can be sent via the intensity channel, which contains information about the intensity value of each point in the point cloud. Next we will explore an example of sending a PointCloud via ROS.

2. Write code to publish PointCloud messages

Publishing PointCloud via ROS is fairly simple. Below we will show a simple example in its entirety and then discuss the important parts of it in detail.

```
1 #include <ros/ros.h>
2 #include <sensor_msgs/PointCloud.h>
3
4 int main(int argc, char** argv) {
5     ros::init(argc, argv, "point_cloud_publisher");
6
7     ros::NodeHandle n;
8     ros::Publisher cloud_pub = n.advertise<sensor_msgs::PointCloud>
      ("cloud", 50);
9
10    unsigned int num_points = 100;
11
12    int count = 0;
13    ros::Rate r(1.0);
14    while(n.ok()){
15        sensor_msgs::PointCloud cloud;
16        cloud.header.stamp = ros::Time::now();
17        cloud.header.frame_id = "sensor_frame";
18
19        cloud.points.resize(num_points);
20
21        //we'll also add an intensity channel to the cloud
22        cloud.channels.resize(1);
23        cloud.channels[0].name = "intensities";
24        cloud.channels[0].values.resize(num_points);
25
26        //generate some fake data for our point cloud
27        for(unsigned int i = 0; i < num_points; ++i){
28            cloud.points[i].x = 1 + count;
29            cloud.points[i].y = 2 + count;
30            cloud.points[i].z = 3 + count;
31            cloud.channels[0].values[i] = 100 + count;
32        }
33
34        cloud_pub.publish(cloud);
35        ++count;
36        r.sleep();
37    }
38 }
```

The important part of this code is explained below.

Line 2: Contains sensor_msgs/PointCloud message headers.

Line 8: Create ros::Publisher, which we will use to send PointCloud messages.

Lines 15–17: Populate the header of the PointCloud message, which we will send with the relevant coordinate system and timestamp information.

Line 19: Set the number of point clouds so we can populate them with dummy data.

Lines 21–24: Add a channel named "intensity" to the point cloud, and resizes it to match the number of points in the cloud.

Lines 26–32: Populate the PointCloud message with some dummy data. At the same time, populate the intensity channel with virtual data.

Line 34: Publish PointCloud messages via ROS.

3. Run to show PointClouds

After writing the sample code, we need to compile it. Before compiling, you need to open the CMakeLists.txt file that was automatically generated when you created the project package and add the following line to the bottom of the file:

```
add_executable(point_cloud_publisher src/ point_cloud_publisher.
cpp)
target_link_libraries(point_cloud_publisher ${catkin_LIBRARIES})
Next, save the file and compile the project package:
$ cd robook_ws
$ catkin_make
```

After compiling the code, we try to run the code. In this part, three terminals need to be opened.

```
The first terminal
$ roscore
Second terminal
Running odometry_publisher:
$ rosrun ch8_setup_tf point_cloud_publisher
Third terminal
Run rviz:
$ rosrun rviz rviz
```

In rviz, you need to make some settings: change fixed frame to sensor_frame; click Add, find PointCloud and add it; change Topic under PointCloud to /cloud. After you finish the above settings, you can see the result as shown in Fig. 8.8, when a bit of cloud appears.

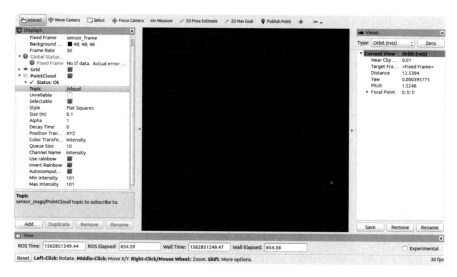

Fig. 8.8 Running results of point_cloud_publisher in rviz

8.5   Configuring and Using the Navigation Project Package Set on Turtlebot

8.5.1   Creating SLAM Maps Via Turtlebot

This section focuses on how to use gmapping to build a map that allows the robot to remember its surroundings. Using the generated map, the robot can achieve autonomous navigation.

First, open a new terminal and start Turtlebot:

```
$ roslaunch turtlebot_bringup minimal.launch
```

Open the second new terminal and start gmapping:.

```
$ roslaunch turtlebot_navigation gmapping_demo.launch
```

Open the third new terminal and start rviz:

```
$ roslaunch turtlebot_rviz_launchers view_navigation.launch
```

Open the fourth new terminal and use the keyboard to control the robot movement:

```
$ roslaunch turtlebot_teleop keyboard_teleop.launch
```

To avoid having to open many terminals each time you build a map, you can combine the procedures needed to build a map into a single map_setup.launch file (available in the book's resource 8.5.1 map_setup.launch) and place it in the launch folder of the project package navigation_test:

```
<! -- Build a map launch file -->
<launch>
  <! --evoke turtlebot -->
  <include file="/home/isi/turtlebot/src/turtlebot/
turtlebot_bringup/ launch/minimal.launch" />
  <! -- Set up the Gmapping package as follows -->
  <! -- Set camera parameters -->
  <include file="/home/isi/turtlebot/src/turtlebot/
turtlebot_bringup/ launch/3dsensor.launch">
    <arg name="rgb_processing" value="false" />
    <arg name="depth_registration" value="false" />
    <arg name="depth_processing" value="false" />
    <arg name="scan_topic" value="/scan" />
  </include>
  <! -- Enabling gmapping -->
  <include file="/home/isi/turtlebot/src/turtlebot_apps/
turtlebot_navigation/launch/includes/gmapping.launch.xml" />
  <! -- Enabling move_base -->
  <include file="/home/isi/turtlebot/src/turtlebot_apps/
turtlebot_navigation/launch/includes/move_base.launch.xml" />
  <! -- Enabling rviz -->
  <include file="$(find turtlebot_rviz_launchers)/launch/
view_navigation.launch" />
</launch>
```

If you want the file to run successfully, you need to replace the path in it with the path to the corresponding file on the user's computer. In most cases, the default navigation parameters provided on turtlebot_navigation can be used, but sometimes the parameters that control the robot's movement need to be modified as appropriate, otherwise they can cause distortions in the created maps. For example, the angular velocity coefficient scale_angular takes a value around 1 and the linear velocity coefficient scale_linear takes a value around 0.2 to build the map better. In addition, the robot should be controlled to try not to have large velocity abrupt changes when building the map, and the distortion is less when the angle of the turn is a multiple of 90°.

Afterwards, save the map. Run the command to save the map in a new terminal window, the path and file name of the save needs to be determined by yourself, the example is as follows:

```
$ rosrun map_server map_saver -f /home/isi/robook_ws/src/
navigation_test/maps/<name>
$ ls /tmp/
```

Now you can see two files my_map.pgm and my_map.yaml in the corresponding locations, which are the maps and parameters to be used for navigation.

8.5.2 Autonomous Navigation Via Turtlebot's Known Maps

Stop the previously open terminal from running. Run the following command in the terminal.

```
$ roslaunch turtlebot_bringup minimal.launch
```

This will launch the Turtlebot robot.
Run the following command from a new terminal.

```
$ roslaunch turtlebot_navigation amcl_demo.launch map_file:=/tmp/
my_map.yaml
```

You can change the directory and name of the map as appropriate. If you see "odom received!", it is working properly.

Run the following command in a new terminal:

```
$ roslaunch turtlebot_rviz_launchers view_navigation.launch --screen
```

rviz can display a map of where the robot is located. The pose of the robot needs to be initialized based on the actual location of the robot.

Select "2D Pose Estimate" on rviz, then click and hold the mouse at the robot's position on the map. At this moment, an arrow will appear under the mouse pointer, which can be used to estimate the direction of the robot. Select "2D Nav Goal" to set the estimated pose at the target position.

This method of taking the target points at runtime each time is cumbersome, and we can get the target's positional parameters programmatically.

Find get_waypoints.cpp and get_waypoints.launch in Sect. 8.5.2 of the book's resources, place them in the /src and /launch folders of the robook/src/ navigation_test project package respectively, and make the relevant CMakeList.txt configuration. In get_waypoints.launch you need to change the directory where the map is actually loaded, e.g.

```
<arg name="map_file" default="/home/isi/robook_ws/src/
navigation_test/ maps/mymap.yaml"/>
```

Modify the location parameters in the get_waypoints.cpp file as appropriate.

```
ofstream posefile("/home/isi/robook_ws/src/navigation_test/src/
waypoints.txt", ios::out);
```

Enter the following command in a new terminal.

```
$ roslaunch navigation_test get_waypoints.launch
```

The initial position of the robot needs to be set by selecting "2D Pose Estimate" on rviz. The robot is controlled by the keyboard to reach the required position, enter get in the small window, and after all points have been acquired enter stop to complete the acquisition of all position points. These parameters can be used in your own navigation program.

Find navigation_demo.cpp and navigation_demo.launch in Sect. 8.5.2 of the book's resources, place them in the /launch and /src folders of the robook/src/ navigation_test package respectively, and configure the CMakeList.txt configuration. You can see the application of the point's bit-pose parameter to the program in the navigation_demo.cpp file. The actual directory where the map is loaded needs to be modified in navigation_demo.launch, e.g.

```
<arg name="map_file" default="/home/isi/robook_ws/src/
navigation_test/ maps/mymap.yaml"/>
```

It can be seen that the target location is obtained by receiving topics in the program, such as shelf, dinner_table and other locations, which can facilitate navigation combined with other functions such as speech recognition.

Run the following command.

```
$ cd robook_ws
$ catkin_make# to compile before first run
$ source devel/setup.bash
$ roslaunch navigation_test navigation_demo.launch
```

Once running, we can determine where the bot is going by sending topics directly via commands. For example.

```
$ rostopic pub /gpsr_srclocation std_msgs/String - shelf
$ rostopic pub /gpsr_dstlocation std_msgs/String - dinner_table
```

Once the topic is successfully posted, the bot will navigate to the appropriate location.

Exercise

Have the robot build a map of the room and navigate to the door via Kinect.

Further Reading

1. Zhang Liang. Research on mobile robot navigation technology in dynamic environment [D]. Wuhan: Wuhan University of Science and Technology, 2013.
2. Peng Zhen. Research on vision-based self-motion estimation and environment modeling methods in dynamic environments [D]. Zhejiang: Zhejiang University, 2017.
3. Shen Jun. Design and implementation of ROS-based autonomous mobile robot system [D]. Mianyang: Southwest University of Science and Technology, 2016.

4. Wang B, Lu W, Kong B. BUILDING MAP AND POSITIONING SYSTEM FOR INDOOR ROBOT BASED ON PLAYER[J]. International Journal of Information Acquisition, 2011, 08 (04):281-290.

5. He Wu, Lu Wei, Wang Yao, et al. Player-based map construction and localization system for indoor service robots[J]. Instrumentation Technology, 2011(5):56-58.

6. Lin Rui. Research on stereo vision SLAM for mobile robots based on image feature points [D]. Harbin: Harbin Institute of Technology, 2011.

7. Morgan Quigley. ROS robot programming practice [M]. Beijing: Mechanical Industry Press, 2018.

8. Ding Linxiang, Tao Weijun. Design and implementation of indoor mobile robot localization and navigation in unknown environment[J]. Military Automation, 2018, 37(3): 12-17.

9. Foxt D, Burgardt W , Thrun S . Controlling synchro-drive robots with the dynamic window approach to collision avoidance[C]. Proceedings of IEEE/RSJ International Conference on Intelligent Robots and Systems. iros 1996, Osaka, Japan, Nov 8-8. 1996.

10. ROS. base_local_planner [EB/OL]. http://wiki.ros.org/base_local_planner.

11. Yang Jingdong. Research on key technologies of autonomous navigation for mobile robots [D]. Harbin: Harbin Institute of Technology, 2008.

12. Li Yuelong. GAP-RBF self-growing self-cancelling neural network on robotics [D]. Chongqing: Chongqing University, 2015.

13. ROS Wiki. Setup and Configuration of the Navigation Stack on a Robot [EB/OL]. http://wiki.ros.org/navigation/Tutorials/RobotSetup.

14. ROS Wiki. Setting up your robot using tf [EB/OL]. http://wiki.ros.org/navigation/Tutorials/RobotSetup/TF.

15. ROS Wiki. publishing Odometry Information over ROS [EB/OL]. http://wiki.ros.org/navigation/Tutorials/RobotSetup/Odom.

16. ROS Wiki. Publishing Sensor Streams Over ROS [EB/OL]. http://wiki.ros.org/navigation/Tutorials/RobotSetup/Sensors.

Chapter 9
Robot Voice Interaction Functions of Basic Theory

Voice is the most friendly and natural way for human-robot interaction. Voice interaction systems for robots include Automatic Speech Recognition (ASR), Semantic Understanding

Speech synthesis (also known as text-to-speech conversion, Text to Speech, TTS) and other basic technologies.

1. Automatic speech recognition technology: it solves the problem of robot "hearing". It is equivalent to the robot's ear, and its goal is to automatically convert human speech content into corresponding text by computer, including signal processing and feature extraction, acoustic model, pronunciation dictionary, language model, decoder and other modules. Here, the choice of microphone is very important and needs to be decided according to the environment of using (whether it is far-field or near-field recognition).
2. Semantic Understanding Technology: It addresses the problem of robot's "understanding". It focuses on how to make robots understand and use natural human language. According to the specific task, it reacts to the recognized speech and sends messages to other nodes. The difficulty of semantic understanding technology is mainly the complexity of the semantics.
3. Speech synthesis technology: converting text to speech output.

This chapter will provide a detailed introduction to the theory underlying the implementation of the above speech interaction features. Among them, the speech recognition part will introduce acoustic models and methods such as Hidden Markov Model, Gaussian Mixture Model, and Deep Neural Network in detail. And language models and methods such as N-gram, NNLM, Word2Vec will be introduced. The semantic understanding section will introduce the Seq2Seq approach in detail. The content of this chapter helps to understand PocketSphinx, the open source speech recognition system used in Chap. 10.This speech recognition system is based on Hidden Markov acoustic model and N-gram language model, and semantic understanding is achieved by executing keyword commands using if conditional judgments.

© The Author(s), under exclusive license to Springer Nature Singapore Pte Ltd. 2023   223
F. Duan et al., *Intelligent Robot*, https://doi.org/10.1007/978-981-19-8253-8_9

9.1   Speech Recognition

As the name suggests, speech recognition is the process of inputting a speech signal and finding a corresponding text sequence that makes the best match between this text sequence and the speech signal. To achieve this, modern speech recognition systems consist of three main core components: an acoustic model, a language model and a decoder. Among them, the acoustic model is mainly used to construct the probabilistic mapping relationship between input speech and output acoustic units; the language model is used to describe the probabilistic collocation relationship between different words, which makes the recognized sentences more like natural text; the decoder is responsible for filtering based on the probability values of acoustic units and the scoring of the language model on different collocations to finally get the most likely recognition results. The speech recognition achieved by acoustic and language models can be expressed as the following equation.

$$\arg\max P(S|O) = \arg\max P(O|S)P(S) \tag{9.1}$$

P(O|S) denotes the acoustic model and P(S) denotes the language model, where O represents the speech input and S represents the text sequence.

9.1.1   Acoustic Models

Traditional acoustic models for speech recognition use the Hidden Markov Model-Gaussian Mixture Model (HMM-GMM) approach, and with the development of deep neural networks, HMM-DNN acoustic models have been commonly used.

9.1.1.1   Hidden Markov Models

The role of an acoustic model is to train a model based on the speech features in the speech database, which is used to construct a probabilistic mapping relationship between the input speech and the output acoustic units, thus enabling acoustic matching of speech information. The acoustic model of the Sphinx speech recognition system used in the case of this chapter is implemented based on the Hidden Markov Model (HMM).

Before introducing the Hidden Markov Model, let's understand Markov Chains. Using a representation of a system with a number of states, which operates for a period of time and undergoes a transfer of states, suppose that the state of the system at the moment is represented by the probability that the state of the system at the moment is related to the state of the system at the moment, represented by the probability

$$P\left(q_t = S_j | q_{t-1} = S_i, q_{t-2} = S_k, \ldots\right) \tag{9.2}$$

The state of the system at moment t is only related to the previous state and not to the previous historical state; a system with this characteristic is called a first-order Markov chain, and this system is discrete, i.e.

$$P\left(q_t = S_j | q_{t-1} = S_i, q_{t-2} = S_k, \ldots\right) = P\left(q_t = S_j | q_{t-1} = S_i\right) \tag{9.3}$$

When considering only the stochastic process associated with moment t, i.e.

$$P\left(q_t = S_j | q_{t-1} = S_i\right) = a_{i,j}, 1 \leq i,j \leq N \tag{9.4}$$

denotes the state transfer probability, which is subject to two conditions: $a_{i,j} \geq 0$ and $\sum_{j=1}^{N} a_{i,j} = 1$.

Since the initial state of the system is not defined, vectors are used to represent the initial probability of each state.

$$\pi = \left\{\pi_i | \pi_i = P(q_t = S_i)\right\} \tag{9.5}$$

$$\sum_{j=1}^{N} \pi_j = 1 \tag{9.6}$$

The hidden Markov model is a statistical model and a twofold Markov stochastic process. "Hidden" indicates that the transfer process between its states is unobservable, which corresponds to the transfer probability matrix; when a state is transferred, it generates or accepts a symbol (i.e., an observed value), which is a stochastic process that corresponds to the emission probability matrix. The Hidden Markov Model can be viewed as a kind of finite state automaton that models the generative process by defining the joint probability of the sequence of observations and the sequence of marks. The difficulty lies in determining the implicit parameters form of the process from the observable parameters.

Given a sequence of observations: $\boldsymbol{O} = O_1, O_2, \ldots, O_T$, the state sequence: $\boldsymbol{Q} = q_1, q_2, \ldots, q_T$, the Hidden Markov Model must satisfy three assumptions:

1. Markov hypothesis: $P(q_i | q_{i-1} \ldots q_1) = P(q_i | q_{i-1})$, the state q_t is only relevant q_{t-1}, independent of the previous state $q_1, q_2, \ldots, q_{t-2}$.
2. Unqualified assumption: for any sum i and j, satisfied $P(q_{i+1} | q_i) = P(q_{j+1} | q_j)$.
3. Output independence assumption: $P(O_1, \ldots, O_T | q_1, \ldots, q_T) = \prod P(O_t | q_t)$

The parameters of a Hidden Markov Model can be expressed as

$$\lambda = (\boldsymbol{N}, \boldsymbol{M}, \boldsymbol{A}, \boldsymbol{B}, \pi) \tag{9.7}$$

In the above equation, $N = \{q_1, \ldots, q_N\}$ represents the states in the model; $M = \{v_1, \ldots, v_M\}$ represents the observed value in the model; $A = \{a_{ij}\}$, $a_{ij} = P(q_t = S_j | q_{t-1} = S_i)$ is the state transfer matrix; $B = \{b_{jk}\}$, $b_{jk} = P(O_t = v_k | q_t = S_j)$ represents the probability emission matrix for the transfer of the implied states to the observed value; and $\pi = \{\pi_i\}$, $\pi_i = P(q_t = S_i)$ is a row vector that represents the probability distribution of the initial states of the model.

Given a sequence of observations: $O = O_1, \ldots, O_T$, the Hidden Markov Model solves the following three main problems:

1. Evaluation problem: Calculate the probability $P(O|\lambda)$ that this observation sequence is generated by the model, which can be solved in practice using a forward-backward algorithm.
2. Decoding problem: How to choose a sequence of states $Q = q_1, q_2, \ldots, q_T$ such that the sequence of observations O is the most probable, i.e., $P(Q|O, \lambda)$ is the maximum value of the solution. In practical applications the Viterbi algorithm can be used to solve it.
3. The learning problem: How to maximize $P(O|\lambda)$ by adjusting the parameters λ, which can be solved in practical applications using the Baum-Welch algorithm.

Hidden Markov models can be applied to speech recognition because the sound signal can be considered as a piecewise stable signal or a short time stable signal and the hidden Markov model has the following characteristics.

1. The hidden condition of moment t is related to the hidden condition of moment t-1. Because human speech has a correlation between before and after, for example, "They are" is often pronounced as "They're", and the pronunciation of sentences is used in speech recognition, so the relationship between the front and back of each syllable needs to be considered in order to The accuracy rate is high. At the same time, the single words in a sentence have relationship context. In terms of English grammar, the subject is often followed by a verb or auxiliary verb. From the usage of single words, the corresponding verb will have a fixed preposition or corresponding noun used, so when analyzing phonological information, in order to improve the accuracy of each single word, it is necessary to analyze the single words before and after.
2. The Hidden Markov Model treats the input information as a signal consisting of a single unit, which is then analyzed, similar to the properties of the human speech model. The unit of speech system recognition is the sound in one unit of time. Using speech processing methods, the sound in a unit of time can be converted into a discrete signal.
3. The hidden conditions used by the Hidden Markov Model are also encapsulated, so it is more appropriate to use the Hidden Markov Model to process sound signals.

9.1.1.2 Gaussian Mixture Model

With the previous introduction of the Hidden Markov Model, we initially understand that the sound input syllable is the observable measure value O in the HMM, and the word corresponding to this sound is the implicit state S of the HMM, and the connection between the implicit state to the observable measure is established through the observation state transfer probability matrix B. The Gaussian Mixture Model (GMM) is used to obtain B, or bi(ot), also known as the likelihood value of the observed state.

First, we use a univariate Gaussian density function to estimate a particular Hidden Markov state i to generate a one-dimensional eigenvector o. Assuming that o satisfies a normal distribution, the likelihood function of the observable variables can be represented by a Gaussian function $b_i(o_t)$, where $\sum_{t=1}^{T} \xi_t(i)$ denotes the number of transfers out of state i

$$b_i(o_t) = \frac{1}{2\pi\sigma_i^2} \exp\left(- \frac{(o_t - \mu_i)^2}{2\sigma_i^2} \right) \tag{9.8}$$

$$\mu_i = \frac{\sum_{t=1}^{T} \xi_t(i)o_t}{\sum_{t=1}^{T} \xi_t(i)} \tag{9.9}$$

$$\sigma_i^2 = \frac{\sum_{t=1}^{T} \xi_t(i)(o_t - \mu_i)}{\sum_{t=1}^{T} \xi_t(i)} \tag{9.10}$$

It is not possible to perform direct speech recognition on a speech signal fed from a microphone, it needs to go through feature extraction. The most commonly used method for speech feature extraction is the Meier Cepstrum Coefficient (MFCC). When a segment of speech signal undergoes feature extraction by Mel inverse spectral coefficients, multi-dimensional features are obtained, and therefore, the univariate Gaussian density function is extended to a multivariate Gaussian distribution for solution. In a Gaussian distribution, it is necessary to assume that each feature is normally distributed, and this assumption is too strong for the actual features, so in practical applications a weighted mixed multivariate Gaussian distribution is often used to solve the likelihood function, i.e., a Gaussian mixture model.

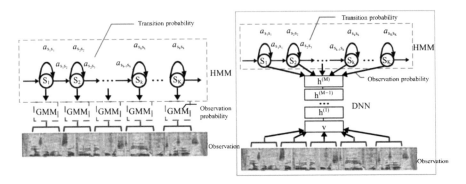

Fig. 9.1 From GMM-HMM to DNN-HMM

$$b_i(o_t) = \sum_{m=1}^{M} c_{jm} \frac{1}{\sqrt{2\pi \mid \sum_{im} \mid}} \exp\left[(x - \mu_{im})^T \sum_{im}^{-1} (o_t - \mu_{im}) \right] \qquad (9.11)$$

From GMM-HMM to DNN-HMM is shown in Fig. 9.1.

9.1.1.3   Deep Neural Networks

Gaussian mixture models can convert speech signal features into a matrix of observed state transfer probabilities that are fed into a Hidden Markov Model for recognition. This can be achieved using a Deep Neutral Network (DNN), which has the following advantages over Gaussian mixture models.

DNNs model the posterior probabilities of speech acoustic features without the assumption of de-distribution of the distribution of the features.

GMM requires the input features to be de-correlated, while DNNs can take various forms of input features.

While GMM can only use a single frame of speech as input, DNN can exploit valid information from the context by stitching together adjacent frames.

Overall, the task of acoustic models is to describe the speech variation patterns and construct probabilistic mapping relationships between input speech and output acoustic units. Acoustic models cover models such as HMM, DNN, RNN, etc. Different acoustic models usually have different recognition performance, while problems such as noise, accent, and articulation habit (swallowing) still exist. If best performance is desired, further optimization is needed in conjunction with the application environment.

9.1.2 Language Models

The language model is a probability-based discriminative model, that is, the input of a sentence (the order of words), the output is the probability of this sentence, that is, the joint probability of these words. For example, in language recognition, for the linguistic input "nixianzaiquganshenme", the first recognition is "what are you rushing to in Xi'an" with P = 0.01; the second recognition is "what are you going to do" with P = 0.2. With the language model, we can get more grammatical results.

9.1.2.1 N-Gram

For a sequence (sentence) $S = (\omega_1, \omega_2, \ldots, \omega_m)$ of m words, each word ω_i is assumed to depend on the influence from the first word ω_1 to the word ω_{i-1} before it, and to calculate the probability $P = (\omega_1, \omega_2, \ldots, \omega_m)$, the chain rule gives:

$$P(S) = P(\omega_1 \omega_2 \ldots \omega_m) = P(\omega_1)P(\omega_2|\omega_1) \ldots P(\omega_m|\omega_{m-1} \ldots \omega_2 \omega_1) \qquad (9.12)$$

However, the above formula has a parameter space that is too large, $p(\omega_m|\omega_{m-1} \ldots \omega_2 \omega_1)$ has $O(m)$ parameters. To solve this problem, the Markov assumption can be used: the current occurrence of a word is related to only a number of words before it. This makes it unnecessary to trace back to the first word, thus drastically reducing the parameter space of the above equation, i.e.

$$P(\omega_i|\omega_1, \omega_2, \ldots, \omega_{i-1}) = P(\omega_i|\omega_{i-n+1}, \ldots, \omega_{i-1}) \qquad (9.13)$$

When N = 1, it is the Uni-gram model, i.e.

$$P(\omega_1, \omega_2, \ldots, \omega_m) = \prod^{m} P(\omega_i) \qquad (9.14)$$

When N = 2, the occurrence of a word at this point depends only on the one word that occurs before it, for the binary model (Bi-gram model).

$$P(S) = P(\omega_1 \omega_2 \ldots \omega_m) = P(\omega_1)P(\omega_2|\omega_1) \ldots P(\omega_m|\omega_m \omega_{m-1}) \qquad (9.15)$$

When N = 3, the occurrence of a word at this point depends only on the two words that occur before it, for the tri-gram model (Tri-gram model).

$$P(S) = P(\omega_1 \omega_2 \ldots \omega_m) = P(\omega_1)P(\omega_2|\omega_1) \ldots P(\omega_m|\omega_m \omega_{m-1}) \qquad (9.16)$$

The order N indicates the complexity and the accuracy of the language model. Practice shows that the larger the value of N, the higher the accuracy of the model, but its complexity is also higher. The value of N is usually taken as 1 to 3. The

Sphinx language model combines a binary grammar with a ternary grammar to calculate the probability of the current word using the probability of the previous one or two words. The probability estimation method calculates the conditional probability by Maximum Likelihood Estimation (MLE) to maximize the probability of the training sample. For the ternary model, the probability of the current word is expressed as.

$$P(\omega_m|\omega_{m-2}\omega_{m-1}) = \frac{P(\omega_{m-2}\omega_{m-1}\omega_m)}{P(\omega_{m-2}\omega_{m-1})} \tag{9.17}$$

Because $P(\omega_{m-2}\omega_{m-1}\omega_m)$ and $P(\omega_{m-2}\omega_{m-1})$ are unknown, the number of $\omega_{m-2}\omega_{m-1}\omega_m$ and $\omega_{m-2}\omega_{m-1}$ needs to be counted from the corpus and denoted as $C(\omega_{m-2}\omega_{m-1}\omega_m)$ and $C(\omega_{m-2}\omega_{m-1})$ respectively. According to the rules of great likelihood estimation, $P(\omega_{m-2}\omega_{m-1}\omega_m)$ can be expressed as

$$P(\omega_{m-2}\omega_{m-1}\omega_m) = \frac{C(\omega_{m-2}\omega_{m-1}\omega_m)}{\sum_{(\omega_{m-2}\omega_{m-1}\omega_m)} C(\omega_{m-2}\omega_{m-1}\omega_m)} \tag{9.18}$$

The above equation represents $C(\omega_{m-2}\omega_{m-1}\omega_m)$ and the ratio to the number of ternary models, and the probability of any N-gram based model can be calculated by the above equation. The number of words contained in the corpus is L, the number of ternary models should be $L-2$, and the number of binary models should be $L-1$. When L is very large, $L-2 \approx L-1$, it is obtained that:

$$P(\omega_m|\omega_{m-2}\omega_{m-1}) = \frac{C(\omega_{m-2}\omega_{m-1}\omega_m)/L-2}{C(\omega_{m-2}\omega_{m-1})/L-1} \approx \frac{C(\omega_{m-2}\omega_{m-1}\omega_m)}{C(\omega_{m-2}\omega_{m-1})} \tag{9.19}$$

However, such an estimate would create a serious problem. If $C(\omega_{m-2}\omega_{m-1}\omega_m)$ is zero, then $P(\omega_m|\omega_{m-2}\omega_{m-1})$is zero and the word sequence will not take into account the ambiguity of the sound signal, giving zero probability for the entire sentence, leading to a data sparsity problem. To address this problem in language models, smoothing techniques are used to adjust the maximum likelihood estimates of probabilities to produce more accurate probabilities. The following smoothing methods are commonly used.

1. Laplace smoothing: force all N-grams to occur at least once by simply adding to the numerator and denominator respectively. The disadvantage of this method is that most of the N-grams do not occur, and it is easy to allocate too much probability space for them.

$$p(\omega_n|\omega_{n-1}) = \frac{C(\omega_{n-1}\omega_n) + 1}{C(\omega_{n-1}) + |V|} \tag{9.20}$$

$C(\omega_{n-1}\omega_n)$ denotes the number of simultaneous and occurrences of ω_{n-1} and ω_n, denotes the number of occurrences of ω_{n-1}, and V denotes the size of the lexicon.

2. Interpolation: the number of possible higher-order combinations is 0, then there are always combinations of lower order points that are not 0. For example, for a third-order combination, suppose that $p(\omega_n|\omega_{n-1}\omega_{n-2}) = 0$, and that p-$(\omega_n|\omega_{n-1}) > 0$, $p(\omega_n) > 0$. Also the weighted average has a probability of not being zero, thus achieving a smoothing effect.

$$\widehat{p}(\omega_n|\omega_{n-1}\omega_{n-2}) = \lambda_3 p(\omega_n|\omega_{n-1}\omega_{n-2}) + \lambda_2 p(\omega_n|\omega_{n-1}) + \lambda_1 p(\omega_n) \qquad (9.21)$$

3. Backtracking: the backtracking method is similar to interpolation, except that it will calculate the probability using the highest order combination possible, and when the higher order combination does not exist, retreat to the lower order until a non-zero combination is found.

$$p(\omega_n|\omega_{n-1}\ldots\omega_{n-N+1})$$
$$= \begin{cases} p^*(\omega_n|\omega_{n-1}\ldots\omega_{n-N+1}) & C(\omega_{n-1}\ldots\omega_{n-N+1}) > 0 \\ \alpha(\omega_{n-1}\ldots\omega_{n-N+1})p(\omega_n|\omega_{n-1}\ldots\omega_{n-N+2}) & otherwise \end{cases}$$
$$(9.22)$$

N-gram language models are very sensitive to the dataset, and the generated language models will be different for different training datasets; therefore, for language models for speech recognition used in daily question and answer for robots, a corpus of questions and answers should be used for training.

9.1.2.2 NNLM

NNLM (Neural Network based Language Model) was proposed by Bengio in 2003. It is a very simple model consisting of an input layer, an embedding layer, an implicit layer and an output layer. The input received by the model is a sequence of words of length n and the output is the next word category, which is shown in Fig. 9.2. First, the input is the index sequence of the word sequence, for example, the index of word I in the dictionary (size |v|) is 10, the index of word am is 23, and the index of Bengio is 65, then the index sequence of the sentence "I am Bengio" is 10, 23, 65. The embedding layer is a matrix of size $|V|{\times}K$, from which the vectors of rows 10, 23, and 65 are taken out to form a $3{\times}K$ matrix, which is the output of the embedding layer. The implicit layer accepts the output of the spliced embedding layer as input, with tanh as the activation function, and finally sends it to the output layer with softmax, where the output is a probability value.

The biggest drawback of NNLM is that it has many parameters and is slow to train. In addition, NNLM requires the input to be of fixed length n. The fact that the input is of fixed length is inflexible, while the complete historical information cannot be utilized.

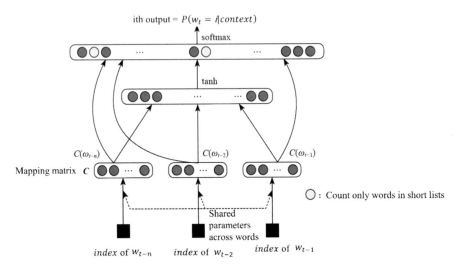

Fig. 9.2 NNLM

9.1.2.3 Word2Vec

Since the beginning of NNLM, the neural networks have been used to build language models, in short , knowing the first k words of the text , use the neural network to predict the probability of the current word. Word2Vec is based on this development. Before we introduce the Word2Vec, we introduce the concept of word vectors.

A word vector is a vector that maps words or phrases from a vocabulary to real numbers. For example, the word king can be represented by the vector [1,0,0,0], queen can be represented by the vector [0,1,0,0], man can be represented by the vector [0,0,1,0], and woman can be represented by the vector [0,0,0,1]. A vector like this used to represent a word is what we call a one-hotencoding of a word vector. Although word vectors are very simple to represent, have many problems, the biggest of which is that vocabularies are generally large. To represent 10 000 words, each word would have to be represented by a vector of length 10 000; to represent 100 000 words, each word would have to be represented by a vector of length 100 000. Thus, the vector used to represent the words is 1 in only one position, and all the rest are 0. The data is sparse and inefficiently expressed. Meanwhile, we can distinguish that king and queen both belong to the royal family, and king and man are both male, so there is some similarity between such words, but the vector of words encoded by one-hotencoding cannot represent this correlation.

To solve these problems, scientists are trained to map each word to a shorter word vector, and all word vectors form the vector space. The relationship between words can be studied by statistical methods, and the dimension of this shorter word vector is specified by the training process. For example, we can represent the above four words ["royal", "sex"] with a two-dimensional word vector representing royalty and gender, then "king" = [0.99,0.99], "queen"=[0.99,0.01], "man"=[0.01,0.99], "woman" = [0.01,0.01]. We call the process of converting the word vector of the one-hotencoding representation to a short continuous vector space representation Word2Vec.

With Word2Vec, we can find that: In the language model of speech recognition, Word2Vec brings better generalization ability. For example.

The cat is walking in the bedroom

A dog was running in a room

If the first sentence has appeared many times in the training corpus, the language model will think that the first sentence conforms to the grammatical rules and the output probability is high; but if the second sentence appears less in the corpus, the language model will think that the second sentence does not conform to the grammatical rules and the output probability is low. But with the Word2Vec method, the language model will know that the and a are similar, and cat and dog are also similar, and give the similar words similar probability, then the second sentence can also get a relatively high output probability.

Language models express the linguistic knowledge contained in natural language by describing the probabilistic collocation relationships between different words. Language models include N-gram, NNLM, Word2Vec, etc. NNLM explores the inherent dependencies in natural language by constructing neural networks, and although NNLM is less interpretable compared to the intuitiveness of statistical language models, this does not prevent it from being a very good way to model probabilistic distributions. Therefore, language models need to be selected and optimized according to specific contexts.

9.2 Semantic Understanding

After the robot recognizes the speech input and converts it into a text sequence, it becomes a key problem that how to get the robot to execute the command according to the result of speech recognition. The simplest way is to execute the keyword commands by using an if-conditional judgment. For example, if the word "follow" is heard, the robot performs the follow action; if the word "face recognition" is heard, the robot performs the face recognition procedure. This is also used in less complex cases.

Here, we will introduce a neural network-based Seq2Seq (Sequence to Sequence) approach that allows robots to "understand" human speech and give answers.

9.2.1 Seq2Seq

First, we briefly introduce recurrent neural networks (RNNs). As shown in Fig. 9.3, where X is the input, U is the weight from the input layer to the implicit layer, s is the

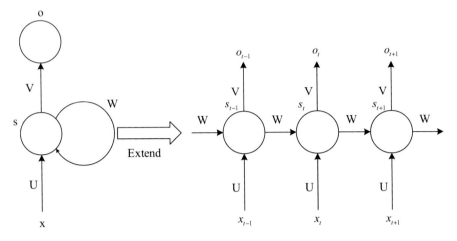

Fig. 9.3 Recurrent neural network structure

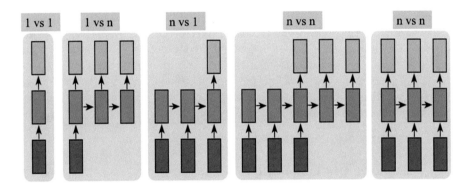

Fig. 9.4 Applications of recurrent neural networks

implicit layer value, W is the weight of the implicit layer at the previous moment as the input at this moment, V is the weight from the implicit layer to the output layer, o is the output, and t is the time variable, t-1 is the previous moment and t+1 is the next moment, with different inputs corresponding to different outputs at different moments.

Depending on the number of inputs/outputs, RNNs can have different structures and different applications(shown in Fig. 9.4)

One-to-one structure: given one input, get one output. This structure does not reflect the characteristics of the sequence, like image classification scenarios.

One-to-many structure: given one input, a series of outputs are obtained. This structure can be used to produce scenes described by pictures.

Many-to-one structure: given a series of inputs, one output is obtained. This structure can be used for text sentiment analysis, such as classifying a series of text inputs and analyzing whether they are negative or positive sentiments.

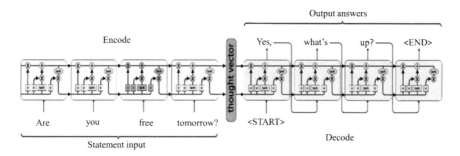

Fig. 9.5   Structure of the Seq2Seq model

Many-to-many structure: give a series of inputs and get a series of outputs. This structure can be used in translation or chat conversation scenarios, where the input text is converted into another series of texts.

Synchronous many-to-many structure: it is the classical RNN structure where the state of the previous input is carried over to the next state and each input corresponds to an output. The most familiar scenario where we apply the synchronous many-to-many structure is character prediction, and it can also be used for video classification and label the frames of a video.

Seq2Seq (shown in Fig. 9.5) belongs to the case of a many-to-many structure in recurrent neural networks, which implements the transformation from one sequence to another. In detail, Seq2Seq belongs to an encoder and decoder structure, and its basic idea is to utilize two RNNs, one RNN as the encoder and the other RNN as the decoder. The encoder is responsible for compressing the input sequence into a vector of specified length, this vector can be seen as the semantics of this sequence, we call it semantic vector, and this process is called encoding. The decoder is then responsible for generating the specified sequence from the semantic vector, a process called decoding, and the simplest way to do this is to input the semantic variables obtained by the encoder as the initial state into the RNN of the decoder to obtain the output sequence, and the output of the previous moment will be used as the input of the current moment. Since each input word in the encoding process has a different impact on each output word in the decoding process, in order to get better decoding results, researchers use an attention mechanism to assign different weights to the input words to generate better output words.

9.3   Speech Synthesis

Speech synthesis, also commonly referred to as Text To Speech (TTS), is a technology that converts any input text into the corresponding speech, and is one of the essential modules in human-computer speech interaction.

Modern TTS processes are very complex. For example, a statistical parametric TTS (SPT) typically has a text front-end for extracting various linguistic features, a duration model, an acoustic feature prediction model, and a vocoder based on complex signal processing. The design of these components requires knowledge from different domains and a lot of effort. They also need to be trained separately, which means that errors from each component can be superimposed on one another. The complexity of modern TTS design requires a great deal of work when building new systems.

Speech synthesis systems usually contain two modules, the front-end and the back-end. The front-end module mainly analyzes the input text and extracts the linguistic information needed by the back-end module. For Chinese synthesis systems, the front-end module usually contains sub-modules such as text regularization, word division, lexical prediction, polysyllabic disambiguation, and rhyme prediction. The back-end module generates speech waveforms by certain methods based on the front-end analysis results. The back-end module is generally divided into two main technical lines: Statistical Parameter Speech Synthesis (SPSS) based on statistical parametric modeling, and speech synthesis based on unit selection and waveform splicing.

Traditional speech synthesis systems generally use Hidden Markov Models for statistical modeling. In recent years, deep neural networks have been increasingly applied to speech synthesis due to their high modeling accuracy. The main neural network models used in speech synthesis techniques are DNN, RNN, LSTM-RNN, etc.

This chapter focuses on the basic theoretical knowledge of robot speech interaction, mainly introducing speech recognition, semantic understanding and speech synthesis technologies. Speech recognition technology solves the problem of "hearing", semantic understanding technology solves the problem of "understanding", and speech synthesis technology is used to convert text into speech output. Through the integration and application of the three basic technologies, the basic speech interaction function of the robot can be realized. The theoretical knowledge of speech technology and acoustic modeling in this chapter can lay the foundation for the operation and concrete implementation of robot speech interaction function in the next chapter.

Further Reading

1. Liu Yi. Grouped gesture recognition based on implicit Markov model [D]. Harbin: Harbin Institute of Technology, 2016.
2. Eddy S R. Hidden Markov models [J]. Current Opinion in Structural Biology, 1996, 6(3): 361-365.
3. Cen Wing-Hua, Han Zhe, Ji Pei-Pei. A study on Chinese term recognition based on hidden Markov model[J]. Data Analysis and Knowledge Discovery, 2008, 24(12): 54-58.
4. Rabiner L R. A Tutorial on Hidden Markov Models and Selected Applications in Speech Recognition [J]. Proceedings of the IEEE, 1989, 77(2): 257-286.
5. Knowing. Notes on speech recognition (IV): a framework for automatic speech recognition based on GMM-HMM [EB/OL]. https://zhuanlan.zhihu.com/p/39390280.

6. CSDN. Acoustic modeling of speech recognition technology [EB/OL]. https://blog.csdn.net/wja8a45tj1xa/article/details/78712930.

7. Leifeng.com. A deep dive into KDDI's latest speech recognition system and framework [EB/OL]. https://www.leip-hone.com/news/201608/4HJoePG2oQfGpoj2.html.

8. CSDN. N-gram models in natural language processing NLP [EB/OL]. https://www.baidu.com/link?url=Usifj7cxHAZ6_H7sUs6fhKAF0dJc1MlauBb07pHTIMVx1jQ_1UvV3sxcKiBKdhHM4wGybtbiOQZ6eHdGCBSl0XDNoUXm6Tw4bpMdYC97W&wd=&eqid=cfead8d70000c2bf000000035cc56699.

9. Cheng, C. Q.. Design and implementation of a shopping guide robot based on speech recognition and text understanding [D]. Wuhan: Wuhan University of Science and Technology, 2014.

10. Li Z, He B, Yu X, et al. Speech interaction of educational robot based on Ekho and Sphinx [C]. Proceedings of the 2017 International Conference on Education and Multimedia Technology, Singapore, July 09–11, 2017, 14–20.

11. CSDN. word2vect basics [EB/OL]. https://blog.csdn.net/yuyang_1992/article/details/83685421.

12. Blogland. Deep learning for seq2seq models [EB/OL]. http://www.cnblogs.com/bonelee/p/8484555.html.

13. CSDN. deep learning of RNN (recurrent neural network) [EB/OL]. https://blog.csdn.net/qq_32241189/article/details/80461635.

Chapter 10
Implementation of Robot Voice Interaction Functionality: PocketSphinx

We have already learned about the basic theory of speech recognition. In terms of concrete implementation, there are many mature packages for speech recognition available, the most closely integrated with ROS are the PocketSphinx speech recognition system and the Festival speech synthesis system. This chapter will introduce PocketSphinx in detail.

First, we introduce the hardware needed for speech recognition, then briefly introduce the PocketSphinx speech recognition system, followed by details on how to install and test PocketSphinx under the Indigo version, and how to publish information about the results of speech recognition via ROS topics to control the robot to perform the appropriate tasks. The installation of PocketSphinx will be different if ROS Kenetic is used, as explained in Sect. 10.4. Speech recognition is an important technology for intelligent service robots and requires proficiency.

10.1 Hardware

Voice is the most friendly and natural way for human-robot interaction. To achieve voice interaction with a robot, the corresponding hardware devices are indispensable. Usually, a microphone is needed as the voice input interface, the processing of voice data through a computer, and an audio or speaker as the output interface for voice. The robot in this book uses an Etus V-Mic D1 microphone (shown in Fig. 10.1) as the voice input interface and a Wanderer M16 audio (shown in Fig. 10.2) as the voice output interface.

© The Author(s), under exclusive license to Springer Nature Singapore Pte Ltd. 2023
F. Duan et al., *Intelligent Robot*, https://doi.org/10.1007/978-981-19-8253-8_10

Fig. 10.1 Microphone

Fig. 10.2 Audio

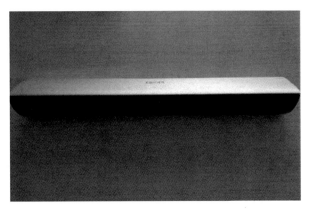

10.2 Introduction of PocketSphinx Speech Recognition System

Sphinx is a continuous speech recognition system developed by Carnegie Mellon University (CMU), USA. In fact, Sphinx is an open source speech recognition package, the current version of which is Sphinx 4.0. It can run on multiple platforms and is portable and scalable and robust. Sphinx supports multiple operating systems such as Windows, Linux, and Unix, and supports recognition of multiple languages. The operating system used in this book is Ubuntu 14.04, and the language recognized is English. As a continuous speech recognition system, Sphinx has a recognition rate of about 80% when the vocabulary is a medium vocabulary and over 98% when the vocabulary is a small vocabulary.

The composition of the Sphinx continuous speech recognition system is shown in Fig. 10.3, with the core components being the acoustic model, the language model, and the speech decoding search algorithm.

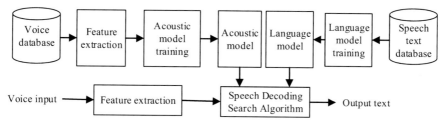

Fig. 10.3 Block diagram of the Sphinx continuous speech recognition system

1. Acoustic model: a model is trained based on the speech features in the speech database, which is used to construct a probabilistic mapping relationship between the input speech and the output acoustic units to achieve acoustic matching of speech information. The acoustic model of the Sphinx speech recognition system is based on the Hidden Markov Model.

2. Language model: grammatical analysis of linguistic texts in the linguistic text database, describing the probabilistic collocation relationships between different words, and using statistical models to build language models so that the recognized sentences are closer to natural texts. Sphinx uses a statistical language model with an N-gram grammatical model (N-gram).

3. Decoder: uses the acoustic model unit probability values with the language model for different pairings to match the characters of the input speech signal with maximum probability and get the most probable recognition result, thus converting the speech signal into text. In this process, a recognition network is created using the dictionary file with the trained language model and acoustic model, and a search algorithm decoder finds an optimal path in the network, this path is a string with maximum probability of recognition of the speech signal. Thus, the decoding operation refers to searching for the best matching string at the decoding end by means of a search algorithm. The current major decoding techniques are based on the Viterbi algorithm, including Sphinx. The Viterbi algorithm is based on dynamic programming, which traverses the hidden Markov state network and retains the best path score for each frame of speech in a given state. The Viterbi algorithm processes the current state information before the next state information when performing state processing. The Viterbi algorithm reduces the scope of the search, increases the speed of the search and is able to achieve real-time synchronization.

The Sphinx speech recognition module consists of the following main parts.

1. Sphinx: a lightweight speech recognition library written in C. It performs the actual decoding operations and generates recognition results using the feature parameters generated in the preprocessing stage and the trained acoustic and language models. It is the key module in a speech recognition system.

2. Sphinx Train: a tool for acoustic model training that uses the eigenvalues from feature extraction for Hidden Markov Model modeling and continuously updates

the variance, weights, mean and other parameters of the model's hybrid Gaussian density function.

3. Sphinx Base: the support library required for Sphinx and Sphinx Train.

Compared to the software development kits provided by Google, Baidu, and KDDI, the most important feature of Sphinx is that it supports offline use, and the methods of offline speech recognition and speech synthesis for robots are briefly described later in this chapter.

10.3   Installing, Testing PocketSphinx on Indigo

10.3.1   Installation of PocketSphinx

1. Download
   Readers can find pocketsphinx-5prealpha.tar.gz and sphinxbase-5prealpha.tar.gz in the book's resources, and also at https://cmusphinx.github.io/wiki/download/.
2. Compile and install
   For compiling and installing PocketSphinx, please refer to https://cmusphinx. github.io/wiki/tutorialpockets-phinx/ or http://wiki.ros.org/pocketsphinx.First unzip the downloaded resources at

```
$ tar -xzf pocketsphinx-5prealpha.tar.gz
$ tar -xzf sphinxbase-5prealpha.tar.gz
```

You should make sure that the following dependencies are installed: gcc, automake, autoconf, libtool, bison, swig (at least version 2.0), the Python development kit, and the pulseaudio development kit. For beginners, it is recommended to install all dependencies. It is recommended to install the dependencies with the command "sudo apt-get install [name]".

Next compile and install Sphinx Base. Change the current directory to the sphinxbase_5prealpha folder and run the following command to generate the configuration file.

```
$ ./autogen.sh
```

The command to compile and install is as follows:

```
$ ./configure
$ make
$ make install
```

The last step may require root access, so you need to run the sudo make install command.

By default, Sphinx Base will be installed in the /usr/local/ directory. Ubuntu will automatically load the libraries from this folder, if not, you will need to configure the

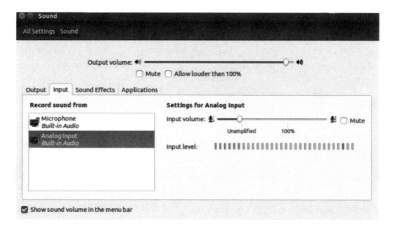

Fig. 10.4 Microphone Test Interface

path to find the shared libraries. This can be done in the /etc/ld.so.conf file, or by exporting the environment variables to:

```
export LD_LIBRARY_PATH=/usr/local/lib
export PKG_CONFIG_PATH=/usr/local/lib/pkgconfig
```

Then, switch to the pocketsphinx folder and perform the same steps.

```
$ ./autogen.sh
$ ./configure
$ make
$ sudo make install
```

3. Testing

After compiling, to test the installation. At this point, run the command pocketsphinx_continuous -inmic yes and check if it recognizes the words you speak through the microphone. In addition, the following installation is required.

```
$ sudo apt-get install ros-indigo-audio-common
$ sudo apt-get install libasound2
```

If errors occurs, such as error while loading shared libraries: libpocketsphinx. so.3, you need to check the linker configuration of the LD_LIBRARY_PATH environment variable above.

If the recognition result is not satisfactory, it may be the microphone setting problem. To set up the microphone device, open the Sound tab in System Settings, select the microphone device to be inserted in Input settings, adjust the input volume, and test the microphone for voice input, as shown in Fig. 10.4.

10.3.2   *Installing a Sound Library for Speech Synthesis*

The Sound_Play package uses the Festival TTS library to generate synthetic speech. First, start the initial Sound_Play node.

```
$ roscore
$ rosrun sound_play soundplay_node.py
```

In another terminal, enter some text to be converted to speech.

```
$ rosrun sound_play say.py "Greetings Humans. take me to your leader."
```

The default voice is called kal_diphone and to see all English voices currently installed on the system.

```
$ ls /usr/share/festival/voices/english
```

To get a list of all available basic Festival sounds, run the following command.

```
$ sudo apt-cache search --names-only festvox-*
```

Some additional sounds can be installed. The following are the steps for obtaining and using two types of sounds, one for men and the other for women.

```
$ sudo apt-get install festlex-cmu
$ cd /usr/share/festival/voices/english/
$ sudo wget -c http://www.speech.cs.cmu.edu/cmu_arctic/packed/
cmu_us_clb_arctic-0.95-release.tar.bz2
$ sudo wget -c http://www.speech.cs.cmu.edu/cmu_arctic/packed/
cmu_us_bdl_arctic-0.95-release.tar.bz2
$ sudo tar jxf cmu_us_clb_arctic-0.95-release.tar.bz2
$ sudo tar jxf cmu_us_bdl_arctic-0.95-release.tar.bz2
$ sudo rm cmu_us_clb_arctic-0.95-release.tar.bz2
$ sudo rm cmu_us_bdl_arctic-0.95-release.tar.bz2
$ sudo ln -s cmu_us_clb_arctic cmu_us_clb_arctic_clunits
$ sudo ln -s cmu_us_bdl_arctic cmu_us_bdl_arctic_clunits
The two sounds can be tested as follows.
$ rosrun sound_play say.py "I am speaking with a female CMU voice"
voice_cmu_us_clb_arctic_clunits
$ rosrun sound_play say.py "I am speaking with a male CMU voice"
voice_cmu_us_bdl_arctic_clunits
```

> Note: If you do not hear the prompt phrase on the first attempt, repeat the command. Also, remember that the soundplay_node node must already be running in another terminal.

10.3.3   *Language Modeling with Online Tools*

1. Create a corpus
   A corpus contains a collection of words, typically sentences, phrases, or words, of speech that needs to be recognized.
   Enter the text of the voice to be recognized in a txt file, e.g.
   stop
   forward
   backward
   Turn right
   turn left

Save and exit, and a simple corpus is created.

2. language modeling with the online tool LMTool
   Go to http://www.speech.cs.cmu.edu/tools/lmtool-new.html and upload the corpus txt file created earlier in the page "Upload a sentence corpus file" to generate. For example, generate TAR3620.tar.gz and unpack it as follows.

   ```
   $ tar xzf TAR3620.tar.gz
   ```

   Of these, the really useful ones are the .dic, .lm suffix files.

3. Test voice library
   The pocketsphinx_continuous decoder uses the -lm option to specify the language model to be loaded and -dict to specify the dictionary to be loaded.

   ```
   $ roscore
   $ rosrun sound_play soundplay_node.py
   $ pocketsphinx_continuous -inmic yes -lm 3620.lm -dict 3620.dic
   ```

At this point, the robot is able to recognize speech, but is not yet able to respond based on the recognition.

4. Send Out the Results of Speech Recognition as a Topic

In order to send the results of the speech recognition as a topic, we use sockets to get the recognition results. The continuous.c under the installation /pocketsphinx-5prealpha/src/programs is to be replaced with 10.2.3 continuous.c from this book's resources and compiled with the gcc continuous.c -c command.

   If you find that the *.h file is missing during compilation, you can find the file in the sphinx-5prealpha directory and copy it under /usr/include/ with administrator privileges. After successful compilation, go to /pocketsphinx-5prealpha under.

```
$ make
$ sudo make install
```

Create a project package named socket, find 10.2.3sever2topic.cpp in the book's resources, and do the CMakeList.txt configuration.

Find 10.2.3speech_demo.py and 10.2.3speech_demo.launch in this book's resources and place them in the src and launch folders inside the speech project package, respectively; find 10.2.3speech_demo.txt, 10.2.3speech_demo. lm, and 10.2.3speech_demo.dic files in the config folder of the speech project package. To give execution rights to the speech_demo.py file.

```
$ cd ~/robook_ws/src/speech/src
$ chmod a+x speech_demo.py
```

Operation.

```
$ cd robook_ws
$ source devel/setup.bash
$ roslaunch speech speech_demo.launch
```

In the new terminal run.

```
$ pocketsphinx_continuous -inmic yes -dict /home/isi/robook_ws/src/
speech/config/speech_demo.dic -lm /home/isi/robook_ws/src/speech/
config/speech_demo.lm
```

At this point, the voice interaction test can be performed based on speech_demo. txt, where the speech_demo.py file is used to process the keyword recognition results, and upper-level changes can be made in this file.

10.4   Installing, Testing PocketSphinx on Kenetic

10.4.1   Installing PocketSphinx on Kenetic versions

First, install the base libraries and components.

```
$ sudo apt-get install ros-kinetic-audio-common libasound2
gstreamer0.10-*
gstreamer1.0-pocketsphinx
Download librshinxbase1_
0.8-6_amd64.deb, install the base shared library. from https://
packages.debian.org/jessie/libsphinxbase1
```

```
$ sudo dpkg -i libsphinxbase1_0.8-6_amd64.deb
Download libpocketsphinx1_0.8-5_amd64.deb, install the front-end
shared libraries.
from https://packages.debian.org/jessie/libpocketsphinx1
$ sudo dpkg -i libpocketsphinx1_0.8-5_amd64.deb
Download gstreamer 0.10-pocketsphinx_0.8-5_amd64.deb, install the
gstreamer plugin from https://packages.debian.org/jessie/
gstreamer0.10-pocketsphinx.
```

```
$ sudo dpkg -i gstreamer0.10-pocketsphinx_0.8-5_amd64.deb
There are so many steps because ROS Kinetic does not support sudo
apt-get install ros-kinetic-pocketsphinx.
Download the American English acoustic model from https://packages.
debian.org/jessie/pocketsphinx-hmm-en-hub4wsj, pocketsphinx-hmm-
en-hub4wsj_0.8-5_all.deb, with the following command.
$ sudo dpkg -i pocketsphinx-hmm-en-hub4wsj_0.8-5_all.deb
```

10.4.2 Testing of PocketSphinx Speech Recognition

Here, we use the open source ROS project on GitHub at

```
$ cd ~/robook_ws/src
```

```
$ git clone https://GitHub.com/mikeferguson/pocketsphinx
```

A new folder for pocketsphinx will be created in our robook_ws/src/ directory, with two subfolders, demo and node. node holds the program for language recognition, and demo holds the launch file for ROS calls, as well as the lexicon for .dic and the language model for .lm. Copy the hub4wsj_sc_8k folder to the pocketsphinx folder, which contains the trained acoustic model.

We next make a change to recognizer.py under robook_ws/src/pocketsphinx/node/.

```
def __init__(self):
  # Start node
  rospy.init_node("recognizer")
  self._device_name_param = "~mic_name" # Find the name of your
  microphone by typing pacmd list-sources in the terminal
  self._lm_param = "~lm"
  self._dic_param = "~dict"
  self._hmm_param = "~hmm" # add hmm parameter
```

```python
def start_recognizer(self):
    rospy.loginfo("Starting recognizer... ")
    self.pipeline = gst.parse_launch(self.launch_config)
    self.asr = self.pipeline.get_by_name('asr')
    self.asr.connect('partial_result', self.asr_partial_result)
    self.asr.connect('result', self.asr_result)
    #self.asr.set_property('configured', True) #mask
    self.asr.set_property('dsratio', 1)


    # Configure language model
    if rospy.has_param(self._lm_param):
        lm = rospy.get_param(self._lm_param)
    else:
        rospy.logerr('Recognizer not started. Please specify a language
        model file.')
        return
    if rospy.has_param(self._dic_param):
        dic = rospy.get_param(self._dic_param)
    else:
        rospy.logerr('Recognizer not started. Please specify a
        dictionary.')
        return
    #from the launch file, get the hmm parameter
    if rospy.has_param(self._hmm_param):
        hmm = rospy.get_param(self._hmm_param)
    else:
        rospy.logerr('Recognizer not started. Please specify a hmm.')
        return
    self.asr.set_property('lm', lm)
    self.asr.set_property('dict', dic)
    self.asr.set_property('hmm', hmm) # Set the hmm parameter
    self.bus = self.pipeline.get_bus()
    self.bus.add_signal_watch()
    self.bus_id = self.bus.connect('message::application', self.
    application_message)
    self.pipeline.set_state(gst.STATE_PLAYING)
    self.started = True
```

Also, we have to modify the launch file under demos by changing robocup.launch
to the following.

```xml
<launch>

<node name="recognizer" pkg="pocketsphinx" type="recognizer.py"
output="screen">
<param name="lm" value="$(find pocketsphinx)/demo/robocup.lm"/>
<param name="dict" value="$(find pocketsphinx)/demo/robocup.dic"/>
<param name="hmm" value="$(find pocketsphinx)/hub4wsj_sc_8k/model/
en-us"/>
</node>
</launch>
```

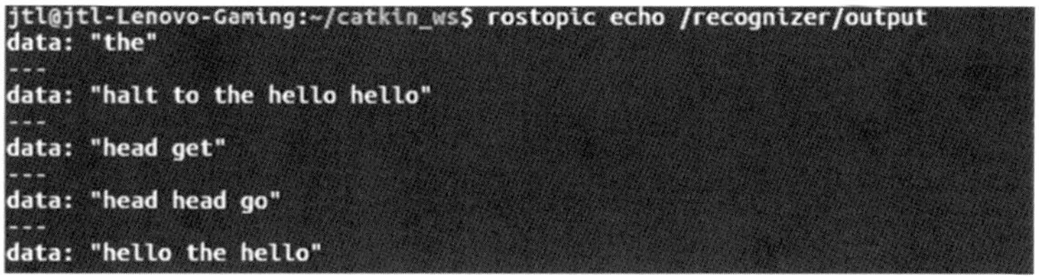

```
jtl@jtl-Lenovo-Gaming:~/catkin_ws$ rostopic echo /recognizer/output
data: "the"
---
data: "halt to the hello hello"
---
data: "head get"
---
data: "head head go"
---
data: "hello the hello"
```

Fig. 10.5 Speech Recognition Results

After modifying the program, we go back to the robook_ws directory under catkin_make, after which we run on a port at

$ roslaunch pocketsphinx robocup.launch

To view the identification results on another port.

$ rostopic echo /recognizer/output

Only words passed into the dictionary (.dic) in the launch file can be recognized here.

The identification results are shown in Fig. 10.5.

1. Voice controlled navigation

Voice control can be used in the virtualizer ArbotiX. We chose the open source package on GitHub for testing (https://github.com/pirobot/rbx1).

First, we run the virtual robot.

$ roslaunch rbx1_bringup fake_pi_robot.launch

Next, open rviz, using the virtualizer profile as a parameter.

$ rosrun rviz rviz -d 'rospack find rbx1_nav'/sim.rviz

Before running the voice recognition script, select the correct input device in "System Settings" → "Sound" and run voice_nav_commands.launch and turtlebot_voice_nav .launch files.

$ roslaunch rbx1_speech voice_nav_commands.launch

In another terminal, run the following command.

$ roslaunch rbx1_speech turtlebot_voice_nav.launch

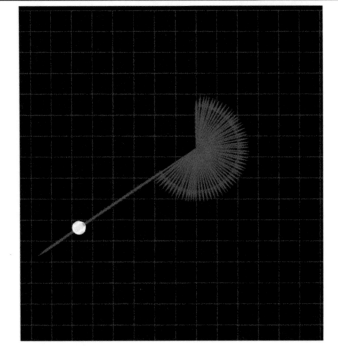

```
process[voice_nav-1]: started with pid [8650]
[INFO] [1551103546.984151]: Ready to receive voice commands
[INFO] [1551103556.523133]: Command: stop
[INFO] [1551103574.885897]: Command: forward
[INFO] [1551103588.893998]: Command: None
[INFO] [1551103592.534712]: Command: stop
[INFO] [1551103597.347861]: Command: rotate right
[INFO] [1551103601.193345]: Command: None
[INFO] [1551103603.205390]: Command: None
[INFO] [1551103605.482777]: Command: stop
[INFO] [1551103609.061052]: Command: stop
[INFO] [1551103611.433746]: Command: forward
[INFO] [1551103619.352729]: Command: None
[INFO] [1551103643.022236]: Command: None
[INFO] [1551103645.393308]: Command: stop
```

Fig. 10.6 Voice Command Recognition Results

At this point, we can try to speak command words to control the virtual robot to move by voice, such as shown in Fig. 10.6.

2. Text to Speech

At this point, the bot is able to recognize what we say. Next, we consider the problem of getting the robot to speak to us as well. The Festival system works with the ROS sound_play package to perform text-to-speech (TTS), i.e. speech synthesis. Again, we have to install the Festival package first (if PocketSphinx speech recognition is already implemented, then you already have the corresponding package installed).

```
$ sudo apt-get install ros-kinetic-audio-common
$ sudo apt-get install libasound2
```

The sound_play package uses the Festival TTS library to generate a mix of voices, and we start by testing the default voice system. First, start the basic sound_play node.

```
$ rosrun sound_play soundplay_node.py
```

Enter the text you want to convert to speech on another terminal. Of course, the Festival system offers other speech styles, which will not be expanded upon here.

```
$ rosrun sound_play say.py "Greetings Humans. take me to your leader."
```

You can also use sound_play to play wave files or self-contained sounds. For example, to play the wave file R2D2 located at rbx1_speech/sounds, use the following command.

```
$ rosrun sound_play play `rospack find rbx1_speech`/sounds/R2D2a.
wavjj
```

To accomplish text-to-speech in the ROS node, we use the talkback.launch file from the rbx1_speech package at

```
$ roslaunch rbx1_speech talkback.launch
```

The startup file first runs the PocketSphinx recognizer node and loads the navigation phrase, the sound_play node is allowed, and then the talkback.py script is run. Now, try saying a voice navigation command, such as "move forward", and you will hear the text-to-speech program output the command word.

To summarize, this chapter first introduce the hardware needed for speech recognition and the PocketSphinx speech recognition system, as well as the installation and testing of PocketSphinx under the Indigo version, and the publishing of speech recognition results through ROS topics to control the robot to perform the appropriate tasks. Finally, the installation of PocketSphinx when using ROS Kenetic is explained. Speech recognition and interaction functions are important technologies in intelligent service robots, which can be used in conjunction with the implementation of navigation functions, vision functions, and robotic arm operation functions of auxiliary robots, and need to be carefully mastered.

Exercises
1. Write a program to give the robot "forward", "backward", "left", "right", etc. by voice. The robot should be able to repeat the commands and move as instructed.

2. Write a program to combine the vision function of the robot, and design the function of using voice to control the robot to call the camera to take pictures, and use the voice to give the robot the command "take pictures", and the robot receives the command and takes pictures with the camera.

Chapter 11
Implementation of Robot Arm Grasping Function

Intelligent service robots often have to help their owners grasp, deliver and transport items, which requires a corresponding robotic arm to do so. The robot in this book uses the Turtlebot-Arm as a robotic arm. This chapter will guide the reader step-by-step on how to use the USB2Dynamixel to control the Turtlebot-Arm robotic arm, starting with the basics of hardware assembly, kinematic analysis, and servo ID settings for this robotic arm, and then detailing how to install and test the dynamixel_motor package under ROS and implement the robotic arm grasping function. This chapter is the basis for implementing the robot to grasp objects in a later chapter and needs to be carefully understood and mastered.

11.1 Components of the Robot Arm Hardware

The robot in this book uses the Turtlebot-Arm as its robotic arm, as shown in Fig. 11.1. The arm consists of five Dynamixel AX-12A servos, with the end 5 servo being the grasping servo, and the arm is controlled by a USB2Dynamixel control board (shown in Fig. 11.2).

For the robotic arm, see the following resources.

A detailed description of the USB2Dynamixel can be found at: http://support. robotis.com/en/product/

auxdevice/interface/usb2dxl_manual.htm.

A description of Turtlebot-Arm's software sources can be found at: http://wiki. ros.org/turtlebot_arm.

The open-source hardware design assembly of the robotic arm can be found at: https://makezine.com/projects/build-an-arm-for-your-turtlebot/.

If using the controller supplied with the ArbotiX, the Turtlebot-Arm robot arm installation software can be found at: http://wiki.ros.org/turtlebot_arm/Tutorials/indigo/Installation.

© The Author(s), under exclusive license to Springer Nature Singapore Pte Ltd. 2023   253
F. Duan et al., *Intelligent Robot*, https://doi.org/10.1007/978-981-19-8253-8_11

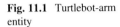

Fig. 11.1 Turtlebot-arm
entity

Fig. 11.2 USB2Dynamixel
control board

If using the controller provided by ArbotiX, the robotic arm servo ID settings can
be found at: http://wiki.ros.org/turtlebot_arm/Tutorials/SettingUpServos.

The Servo ID settings can be found at: http://wiki.ros.org/dynamixel_controllers/
Tutorials/SettingUpDynamixel. The Servo ID settings can also be found in the next
Sect. 11.3.

11.2   Kinematic Analysis of the Robot Arm

Robotic arms grasp items involving positive and negative kinematics problems. The
kinematic control varies depending on the composition of the robotic arm, but the
kinematic analysis is much the same. The coordinate system of the robotic arm used
in this book is shown in Fig. 11.3. In this book, to ensure stable grasping, the joints
of the robotic arm should be made horizontal in the grasping state. Let the

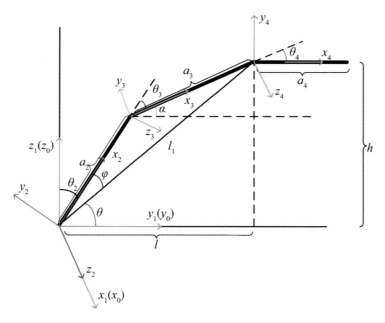

Fig. 11.3 Coordinate system of robot arm

coordinates of the object with respect to the polar coordinate system of the robotic arm is ($p.\,x, p.\,y, p.\,z$), considered $h = p.\,z$.

Using the geometric method of solving, the following equation is obtained from the robot arm coordinate system.

$$\begin{cases} l = \sqrt{p.x^2 + p.y^2} - a_4 \\ a_2 \cos\theta_2 + a_3 \cos\theta_3 = l \\ a_2 \sin\theta_2 + a_3 \sin\theta_3 = p.z \\ l_1 = \sqrt{l^2 + p.z^2} \\ \cos\varphi = \dfrac{l_1^2 + a_2^2 - a_3^2}{2a_2 l_1} \\ \theta_1 = -\arctan(p.x, \ p.y) \\ \theta_2 + \theta_3 + \theta_4 = 90° \\ \theta_2 = 90° - \arctan\dfrac{p.z}{l} - \varphi \end{cases} \qquad (11.1)$$

The equation for each joint angle is as follows

$$\begin{cases} \theta_1 = -\arctan(p.x,\ p.y) \\ \theta_2 = 90\,° - \arctan\dfrac{p.z}{l} - \varphi \\ \theta_4 = \theta_2 - \arccos\left(\dfrac{l^2 + p.z^2}{2a_2a_3} - 1\right) \\ \theta_3 = 90\,° - \theta_2 - \theta_4 \end{cases} \tag{11.2}$$

The angle θ_1, θ_2, θ_3, θ_4 at which the servo needs to be turned, θ_5 is the gripping angle of the gripper. The user can adjust the general arc $\theta_5 = 0.2 - 0.3$ as needed.

11.3 Setting the Servo ID of the Robot Arm

The default ID of each Dynamixel AX-12A servo of the robot arm is 1. The ID of the servo is set before controlling the robot arm.

This section describes a method of modifying the Servo ID using ROBOPLUS software and using USB2Dynamixel under Windows. The relevant preparations are as follows.

System: Windows.

Software: ROBOPLUS official software.

Hardware: USB2Dynamixel serial module, SMPS2Dynamixel power supply module, 12V5A adapter, AX-12A servo (or other servo models)

Download RoboPlus (depending on system choice) from the official website at http://en.robotis.com/service/

downloadpage.php?ca_id=1080; or use RoboPlusWeb (v1.1.), which is attached to this book.

3.0).exe and install it.

Plug the USB2Dynamixel serial module into the USB port of your computer and the driver for the USB2Dynamixel serial module will be installed automatically. Follow Fig. 11.4 for hardware connection.

> Note: Be sure the microswitch on the side of the USB2Dynamixel controller has been moved to the correct setting first. For 3-pin AX-12, AX-18 or new T-series servos (e.g. MX-28T), TTL needs to be set. for 4-pin or R-series servos (e.g. MX-28R, RX-28, EX-106+), RS-485 needs to be set. once the connection is done correctly, the red LED on the controller will light up.
> Open the Device Manager on your computer and check the serial port number of the USB2Dynamixel serial module, as shown in Fig. 11.5.

Open RoboPlus again and select "Dynamixel Wizard" under "Expert Edition", as shown in Fig. 11.6.

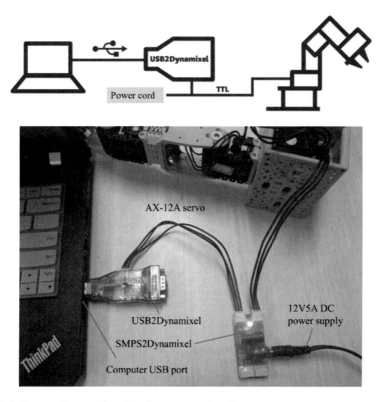

Fig. 11.4 Servo and control board hardware connection diagram

In the pop-up dialog box, select the serial port number (here COM4 as previously viewed) and click the Connect button to make the connection between the software and the USB2Dynamixel, as shown in Fig. 11.7.

After a successful connection, select the baud rate and begin searching for a servo, as shown in Fig. 11.8. For new servos, the baud rate used is typically 1 000 000 for the AX series servos and 57142 for the MX and RX series.

The search results are shown in Fig. 11.9, which shows that five servos were detected with IDs 1, 2, 3, 4, and 5. To view the details of a particular servo or to modify the ID, click on the corresponding servo.

Note: If the servo is new, the servo ID is usually 1 or 0. This is not the time to modify the ID as in Fig. 11.9. Because the servo IDs are all the same, this method will result in the servo not being detected eventually. If the servos are all new, it is recommended that you modify the servo IDs individually before using them.

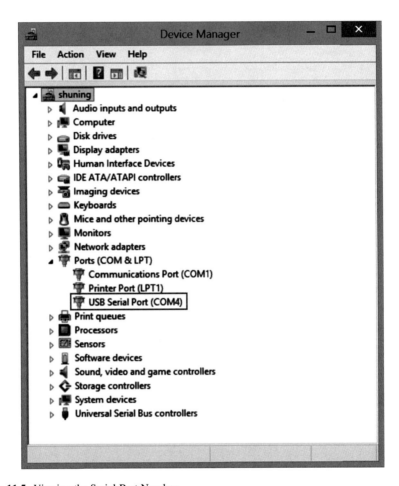

Fig. 11.5 Viewing the Serial Port Number

Clicking on a detected servo displays the details of that servo, as shown in Fig. 11.10.

At column 1 above, the baud rate, ID number, and servo model of the servo are displayed; in column 2, all the current information of the servo is displayed; and in column 3, the mode of the servo is displayed.

Now let's see how to change the Servo ID number. As shown in Fig. 11.11.

First, select the servo to be modified, then check the ID number column, then click the drop-down menu to bring up the available ID numbers and select the ID number to be modified; finally, click the Apply button. In this way, the ID number is modified.

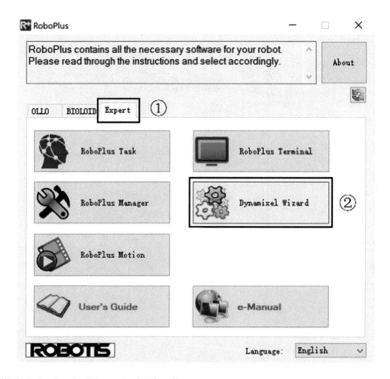

Fig. 11.6   Selecting the "Dynamixel Wizard"

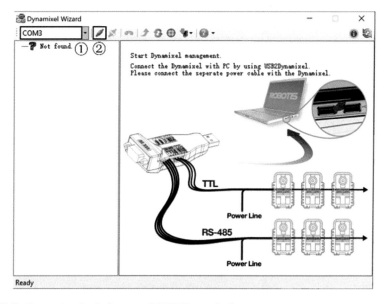

Fig. 11.7   Connecting the Software and USB2Dynamixel

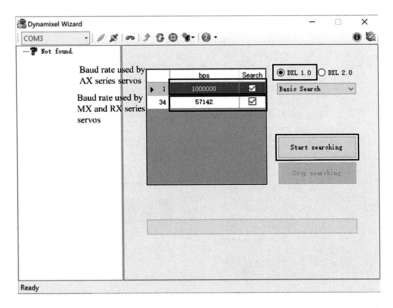

Fig. 11.8 Searching for a Servo

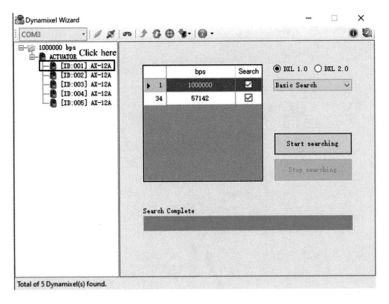

Fig. 11.9 Search Results for Servo

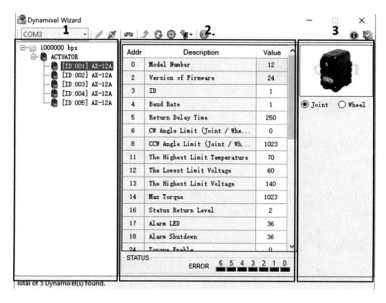

Fig. 11.10   Viewing Servo Information

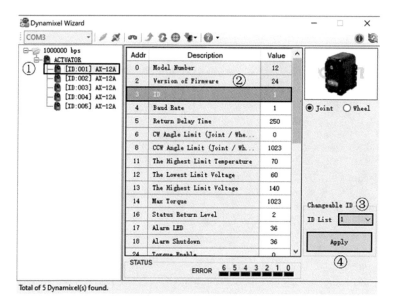

Fig. 11.11   Modifying the Servo Number

11.4   Controlling Turtlebot-Arm with USB2Dynamixel

The Ubuntu package corresponding to USB2Dynamixel is available from the ROS Ubuntu package repository. You can refer to http://wiki.ros.org/dynamixel_motor and http://wiki.ros.org/dynamixel_controllers/Tutorials.

11.4.1   Installing and Testing the dynamixel_motor Package

1. Install the package
   To install dynamixel_motor and its included packages, simply execute the following command.

   ```
   $ sudo apt-get install ros-%ROS_DISTRO%-dynamixel-motor
   ```

   For example.

   ```
   $ sudo apt-get install ros-indigo-dynamixel-motor
   ```

   However, we commonly use a different installation method. First, copy the program files.

   ```
   $ cd ~/robook_ws/src
   $ git clone https://github.com/arebgun/dynamixel_motor.git
   ```

   Then compile the files.

   ```
   $ cd ~/robook _ws
   $ catkin_make
   ```

   Create your own robotic arm engineering package.

   ```
   $ cd ~/robook_ws/src
   $ catkin_create_pkg ch11_dynamixel dynamixel_controllers std_msgs
   rospy roscpp
   ```

2. Test hardware connection
   Follow the connection hardware diagram given in Sect. 11.2 to make the connections. Assume that the USB2Dynamixel is connected to the /dev/ttyusb0 serial port. You can see which USB port the USB2Dynamixel is connected to by entering the following command.

   ```
   $ ls /dev/ttyUSB*
   ```

   Normally, you would see output like the following.

```
/dev/ttyUSB0
```

If you see the message as shown below.

```
ls: cannot access /dev/ttyusb* No such file or directory
```

Then your USB2Dynamixel is not being recognized. At this point, try plugging it into a different USB port, use a different cable, or check your USB hub. If no other USB devices are connected, then the USB2Dynamixel will be located under the directory /dev/ttyUSB0. If other USB devices need to be connected at the same time, insert the USB2Dynamixel first so that it can be assigned to device /dev/ttyUSB0.

If there are other problems, you should check to see if you need to add the user account to the dial-out group and if you need to set the USB port to writable. Sometimes there are errors such as.

could not open port/dev/ttyUSB0: [Errno 13] Permission denied: /dev/ttyUSB0

This is done by executing the following command.

```
$ sudo dmesg -c
$ sudo chmod 666 /dev/ttyUSB0
```

3. Start the Controller Manager

We need to start the controller manager, connect to the servo at the specified rate and post raw feedback data (e.g. current position, target position, errors, etc.). The easiest way to do this is to write a launch file that will set all the necessary parameters. The following text needs to be copied and pasted into the ~/robook_ws/src/ch11_dynamixel/launch/ controller_manager.launch file.

```
    <! -- -- mode: XML -- -->
<launch>
  <node name="dynamixel_manager" pkg="dynamixel_controllers"
type="controller_manager.py" required="true" output="screen">
    <rosparam>
      namespace: dxl_manager
      serial_ports:
        pan_tilt_port:
          port_name: "/dev/ttyUSB0"
          baud_rate: 1000000
          min_motor_id: 1
          max_motor_id: 25
          update_rate: 20
    </rosparam>
  </node>
</launch>
```

Note: You should ensure that the baud rate is set correctly. The AX-12 servo is used here with a baud rate of 1000000. if the RX-28 servo is used, it should be set to 57142.
If USB2Dynamixel is assigned to device /dev/tty USB1, all of the above dev/tty USB0 should be modified to dev/tty USB1.

Run controller_manager.launch.

```
$ cd robook_ws
$ source devel/setup.bash
$ roslaunch ch11_dynamixel controller_manager.launch
```

The ID search defaults from 1 to 25. If it does not find it, you can change controller_manager.launch to expand the ID search.

Right now, the controller manager is posting topics about /motor_states/pan_tilt_port. First, check to see if the topic exists.

```
$ rostopic list
```

The output should be similar to that shown below.

```
/motor_states/pan_tilt_port
/rosout
/rosout_agg
```

4. Specify controller parameters

First, we need to create a configuration file that contains all the parameters needed for the controller. Paste the following text into the tilt.yaml file, which can be placed in the config folder under the ch11_dynamixel project package.

```
tilt_controller:
controller:
package: dynamixel_controllers
module: joint_position_controller
type: JointPositionController
joint_name: tilt_joint
joint_speed: 1.17
motor:
id: 1
init: 512
min: 0
max: 1023
```

Note: You should make sure that the motor ID matches the servo ID used.
motor section has four parameters: id, init, min and max.
The id parameter should be the same as the id of the servo used.
The init parameter has 512 values and it varies between 0 and 1023. This
  parameter is related to the initial position of the joint. Since the full rotation
  is 360 degrees, set up the init: 512 will keep the initial state of the servo
  120 degrees from the original reference 0.
The min parameter is the minimum rotation the servo can do, and it follows the
  same rules as the init parameter.
The max parameter is the maximum rotation the servo can do, and it follows
  the same rules as the parameters above.

5. Create the launch file

   Next, we need to create a launch file that will load the controller parameters to
   the parameter server and start the controller. Paste the following text into the
   start_tilt_controller.launch file.

```
<launch>
  <! -- Start tilt joint controller -->
  <rosparam file="$(find my_dynamixel_tutorial)/tilt.yaml"
  command="load"/>
  <node name="tilt_controller_spawner" pkg="dynamixel_controllers"
  type="controller_spawner.py"
    args="--manager=dxl_manager
          --port pan_tilt_port
          tilt_controller"
    output="screen"/>
</launch>
```

6. Start controller

   Starting the controller will run the file created above. We start the Controller
   Manager node first by entering the following command as in step 3 above.

```
$ cd robook_ws
$ source devel/setup.bash
$ roslaunch ch11_dynamixel controller_manager.launch
Once the Controller Manager is up and running, we can load the
controller:
$ cd robook_ws
$ source devel/setup.bash
$ roslaunch ch11_dynamixel start_tilt_controller.launch
```

The output is similar to the display below. If everything starts properly, you will
see "Controller tilt_controller successfully started" in the terminal.

```
process[tilt_controller_spawner-1]: started with pid [4567]
[INFO] 1295304638.205076: ttyUSB0 controller_spawner: waiting for
controller_manager to startup in global namespace...
[INFO] 1295304638.217088: ttyUSB0 controller_spawner: All services
are up, spawning controllers...
[INFO] 1295304638.345325: Controller tilt_controller successfully
started.
[tilt_controller_spawner-1] process has finished cleanly.
```

Next, we list the topics and services offered by Dynamixel controllers.

```
$ rostopic list
Related topics are as follows.
/motor_states/ttyUSB0
/tilt_controller/command
/tilt_controller/state
The /tilt_controller/command topic requires a message of type
std_msgs/float64 for setting the joint angle.
The /tilt_controller/state topic provides the current state of the
servo, with message type dynamixel_msgs/JointState.
The relevant services are as follows.
/restart_controller/ttyUSB0
/start_controller/ttyUSB0
/stop_controller/ttyUSB0
/tilt_controller/set_compliance_margin
/tilt_controller/set_compliance_punch
/tilt_controller/set_compliance_slope
/tilt_controller/set_speed
/tilt_controller/set_torque_limit
/tilt_controller/torque_enable
```

Here are some services for changing servo parameters such as speed, motor torque limit, flexibility, etc.

7. Control the rotation of the servo

   To make the servo turn, we need to post the desired angle to the / tilt_controller/command topic, as follows.

```
$ rostopic pub -1 /tilt_controller/command std_msgs/Float64 -- 1.5
```

At this point you can see the rudder turning 1.5 radians.

11.4.2 Implementation of the Robotic arm Gripping Function

Robotic arms are mostly used for grasping, and in addition to the basics presented in the previous section, the robotic arm part involves inverse kinematics and has to be

combined with other functions such as vision, which will be explained step by step
next.

1. Specify controller parameters

    First, we need to create a configuration file that contains all the parameters
needed for the controller. Paste the following text into the joints.yaml file, which
can be saved in the config folder under the ch11_dynamixel project package.

```
joints: ['arm_shoulder_pan_joint', 'arm_shoulder_lift_joint',
'arm_elbow_flex_joint', 'arm_wrist_flex_joint', 'gripper_joint']
arm_shoulder_pan_joint:
  controller:
    package: dynamixel_controllers
    module: joint_position_controller
    type: JointPositionController
  joint_name: arm_shoulder_pan_joint
  joint_speed: 0.75
  motor:
    id: 1
    init: 512
    min: 0
    max: 1024
arm_shoulder_lift_joint:
  controller:
    package: dynamixel_controllers
    module: joint_position_controller
    type: JointPositionController
  joint_name: arm_shoulder_lift_joint
  joint_speed: 0.75
  motor:
    id: 2
    init: 512
    min: 0
    max: 1024
arm_elbow_flex_joint:
  controller:
    package: dynamixel_controllers
    module: joint_position_controller
    type: JointPositionController
  joint_name: arm_elbow_flex_joint
  joint_speed: 0.75
  motor:
    id: 3
    init: 512
    min: 0
    max: 1024
arm_wrist_flex_joint:
  controller:
    package: dynamixel_controllers
    module: joint_position_controller
    type: JointPositionController
  joint_name: arm_wrist_flex_joint
```

```
joint_speed: 0.75
motor:
   id: 4
   init: 512
   min: 0
   max: 1024
gripper_joint:
 controller:
   package: dynamixel_controllers
   module: joint_position_controller
   type: JointPositionController
 joint_name: gripper_joint
 joint_speed: 0.35
 motor:
   id: 5
   init: 512
   min: 0
   max: 1024
```

Note: Be sure that each motor ID matches the servo ID used and that the servos are all operating in series mode.

2. Inverse kinematic control of joint angles

Create the arm_grasp.py file in the src folder under the ch11_dynamixel project package, copy the contents of 11.4.2arm_grasp.py from the book's resources to that file, and save it. Note that to give execution rights to the file.

```
$ cd ~/robook_ws/src/ch11_dynamixel/src
$ chmod a+x arm_grasp.py
The relevant code for the inverse kinematic operations in the file is as
follows.
### Inverse kinematic operations
# calculate new joint angles
self.theta1 = - math.atan2(self.x,self.y) + 0.15
#because the arm changed to the new one, "+ 0.05 changed to "-0.05"
self.l = math.hypot(self.x,self.y) - self.a4
self.l1 = math.hypot(self.l,self.z)
print self.l1
self.cosfai = self.l1 / (2 * self.a2)
rospy.loginfo("cosfai is ....") #print x coordinate
print self.cosfai
if self.cosfai > 1:
    self.cosfai = 1
self.fai = math.acos(self.cosfai)
self.theta = math.atan2(self.z,self.l)
self.theta2 = math.pi / 2 - self.theta - self.fai
self.theta4 = self.theta - self.fai
self.theta3 = 2 * self.fai
```

```
#self.theta4 = self.theta2 - math.acos(self.l1 ** 2 / (2 * self.a2 *
self.a3) - 1 )
#self.theta3 = math.pi / 2 - self.theta2 - self.theta4
```

3. Create the launch file

Next, we need to create a launch file that will load the controller parameters to the parameter server and launch the controller. Paste the following text into the arm_grasp.launch file.

```
<launch>
 <node name="dynamixel_manager" pkg="dynamixel_controllers"
 type="controller_manager.py" required="true" output="screen">
 <rosparam>
      namespace: dxl_manager
      serial_ports:
        servo_joints_port:
          port_name: "/dev/ttyUSB0"
          baud_rate: 1000000
          min_motor_id: 1
          max_motor_id: 5
          update_rate: 20
   </rosparam>
 </node>
 <rosparam file="$(find ch11_dynamixel)/config/joints.yaml"
 command="load"/>
 <node name="controller_spawner" pkg="dynamixel_controllers"
 type="controller_spawner.py"
    args="--manager=dxl_manager
        --port servo_joints_port
        arm_shoulder_pan_joint
        arm_shoulder_lift_joint
        arm_elbow_flex_joint
        arm_wrist_flex_joint
        gripper_joint"
    output="screen"/>
 <node name="arm_grasp" pkg="ch11_dynamixel" type="arm_grasp.py"
 output="screen"/>
</launch>
```

4. Running

The Turtlebot robot and Primesense camera were connected separately to a laptop, and the robotic arm was connected using a USB2Dynamixel controller, and the Primesense camera was used for target identification and location, and the robot was gradually moved to the working space of the robotic arm by adjusting its position in order to realize the grasping of objects by the robotic arm.

Refer to Sect. 7.3.3 for object identification and localization, with the following code.

```
$ cd robook_ws
$ catkin_make# to compile before first run
$ source devel/setup.bash
$ roslaunch imgpcl objDetect.launch
```

The code for the robotic arm control is as follows.

```
$ cd robook_ws
$ source devel/setup.bash
$ sudo dmesg -c
$ sudo chmod 666 /dev/ttyUSB0
$ roslaunch ch11_dynamixel arm_grasp.launch
Identify the name of the target object by topic posting, with the
following code.
$ rostopic pub objName std_msgs/String -- potatoChips
```

After a successful run, the robotic arm is able to grasp the target item.

Exercise
Programmed to control the robot to grab the mineral water bottle on the table.

Part III
Applications of Robots

The intelligent home service robot designed in this book has multiple functions, and although its functions can be realized individually and stably, they are far from being able to cope with the complexity of the home environment and family life, and the functions need to cooperate with each other to provide more friendly and intelligent services. To demonstrate the comprehensive performance of the robot, we designed three integrated cases in the home environment to realize the functions of long command recognition and multitask execution, following and assisting the owner, and customer waving to indicate the robot to order food, etc.

In order to simulate a real home environment, this book uses a laboratory and a hallway to build a home environment and arrange some furniture such as a table, sofa, coffee table, etc. We try to simulate a real home environment and test the various performances of the robot in this environment. Through this part, the reader can understand how to implement a complete intelligent service robot by combining the functions of autonomous navigation, visual recognition and voice recognition introduced earlier. Through these cases, the reader can also get a complete overview and deeper understanding of what has been learned earlier.

It should be noted that the program in this section is implemented on Ubuntu 14.04, Indigo version, and we do not elaborate on the program code; the program can be viewed in the resources of this book.

Chapter 12
Integrated Robots Application Case 1: Long Command Recognition and Multitasking Execution

One of the basic capabilities of an intelligent service robot is to interact with its owner, recognize the owner's commands, and execute the corresponding tasks as ordered. The most natural way to interact is for the owner to give explicit commands according to natural language conventions, and for the robot to understand the commands and execute them correctly. Usually, a complete linguistic command includes information such as location, object, color, and person, which are combined into a long voice command. The robot decomposes this long voice command into understandable keywords through voice recognition technology, and creates multiple subtasks which correspond to the corresponding keywords, and then executes them in the order of the tasks to complete the user's entire command finally.

12.1 Case Objectives

This case will allow a home service robot to perform a common task in family life. The goal of the task is that the robot recognizes the voice command given by the user and subsequently goes to a certain place to retrieve the required item as commanded. The specific process is that the user gives the robot a voice command to execute the task of going to a certain place to pick up a certain item. At this point, the robot has to accurately recognize the user's command, analyze the requested destination and the item to be fetched in the voice command, automatically navigate to the vicinity of the destination, and use the robot arm to grab the item. This task mainly uses robot voice recognition, autonomous navigation, and obstacle avoidance, object recognition and grasping, and other functions.

Next, we will analyze an example. The user gives the voice command "Kamerider, go to the table, find green tea, and give it to the person in the sofa". Here, "Kamerider" is the name of the service robot designed in this book. First, the robot recognizes the user's voice commands, analyzes the destination "table", "sofa" and the item "green tea", and then the robot navigates autonomously to "table",

© The Author(s), under exclusive license to Springer Nature Singapore Pte Ltd. 2023
F. Duan et al., *Intelligent Robot*, https://doi.org/10.1007/978-981-19-8253-8_12

where it activates its image processing and robot arm grasping functions (vision servo), and detects the number of objects. After successfully grasping the object ("green tea"), the robot navigates again to another location ("sofa"). At this point, the basic task of this case is complete, but we can refine the design to allow it to perform a series of more advanced tasks: after arriving at the "sofa", the robot can again be controlled by voice, such as recognizing the command "Kamerider, Okay" and then drop the object; it can communicate with the robot through voice questions and answers, such as asking for the number of objects detected. The entire workflow is roughly shown in Fig. 12.1.

In the following, the example will be analyzed in terms of voice recognition, autonomous navigation, and object recognition and grasping.

12.2 Voice Recognition Tasks

To make a home service robot obey the user's commands, it is first necessary to build a thesaurus beforehand, based on the robot's tasks.

Such as kamerider, sofa, in-the-sofa, table, to-the-table, shelf, to-the-shegreen-tea, potato-chips, cola, okay, how-many, person, time, how-old, question. Among them, place nouns are sofa, table, and shelf, and object nouns are green-tea, potato-chips, and cola. The robot should extract the keywords from the recognized commands, assign the keywords to different tasks, and send the keywords to the corresponding topics. When the robot hears the command "Kamerider, go to the table find the green tea and give it to the person in the sofa", it needs to extract "table", "green-tea", "person", "sofa", etc. and send it to the corresponding topic. The navigation part will subscribe to the topic and get the keywords "table" and "sofa" for the location to be visited; the image part will learn that the object to be recognized is "green-tea"; the robot arm part will learn that the object to be recognized is "green-tea". "; the robot arm will learn that the object to be grasped is "green-tea" after the grasping is completed the navigation system will go to the second location "sofa"; after the recognition of " Kamerider" and "Okay" is completed the robot arm will drop the object.

The analysis block diagram for voice recognition is shown in Fig. 12.2.

12.3 Autonomous Navigation in the Home Environment

For a home service robot to provide good service to the user, the robot must be made familiar with the entire home environment, so first, a map of the home environment needs to be established for the robot to navigate autonomously on the known map. Referring to Sect. 8.5, a map of the home environment is created to control the robot's movement through the home environment. Fig. 12.3 shows a map of the

Fig. 12.1 Case Flow Chart

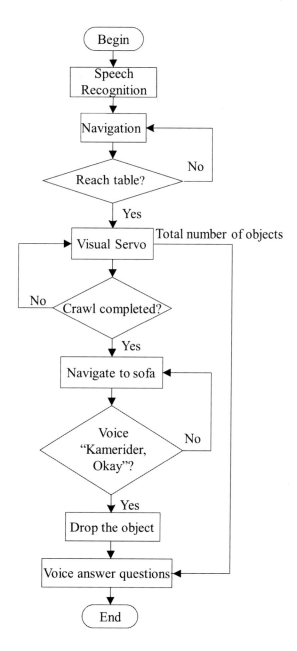

simulated home environment with furniture such as sofas and coffee tables labeled in the figure.

In practical applications, the robot should be able to smoothly avoid various obstacles that appear during navigation so that it can better serve people. In this example, to test the robot's obstacle avoidance function, in addition to various

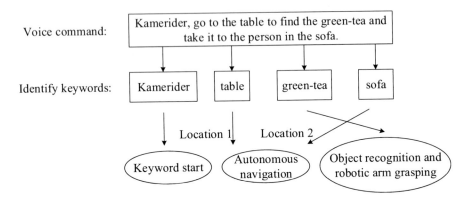

Fig. 12.2 Block diagram of voice recognition analysis

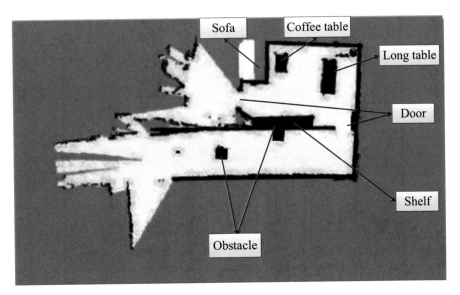

Fig. 12.3 Map of the simulated home environment

household obstacles, we added two obstacles to the robot's forward path and tested its ability to avoid dynamic obstacles (when someone suddenly appeared in front of the robot, the robot could avoid them smoothly). The results show that the robot has good autonomous navigation, autonomous obstacle avoidance, and autonomous path planning. The robot in Fig. 12.4a was able to avoid obstacles while planning its path without running into them; the robot in Fig. 12.4b was able to pass through a narrow doorway smoothly, this is due to some parameters set in navigation; the robot in Fig. 12.4c reached the "table" position and stopped, ready to grab the item; the robot in Fig. 12.4d navigates to the "sofa" after grabbing the item and hands the item "green tea" to the user.

|a) Avoid obstacles|b) Pass through the narrow doorway smoothly|

|c) Reach the "table" position|d) Reach the "sofa" position|

Fig. 12.4 Robot autonomous navigation, obstacle avoidance, and path planning process. (**a**) Avoid obstacles. (**b**) Pass through the narrow doorway smoothly. (**c**) Reach the "table" position. (**d**) Reach the "sofa" position

12.4   Object Recognition and Grasping

When the robot receives the grasp command and identifies the information of the object to be grasped, the Primesense RGBD vision sensor collects the color information and depth information of the environment, detects the 2D position of the object in the color image using the sliding window template matching method based on the Hue histogram, and then gets the corresponding spatial 3D position of the object according to the 2D position, and judges according to the 3D position of the object Whether the object is within the grasping range of the robotic arm. When the object is outside the working space of the robot arm, the robot fine-tunes its position according to the object's position relative to the robot, and then repeats the image recognition and positioning process until the object is within the working space of the robot arm and the robot arm performs inverse kinematic grasping according to the 3D position of the object. The process of the robot recognizing and grasping the object is shown in Fig. 12.5.

Exercise
Design different objects and scenarios to test the robot's navigation and grasping success rate under different room layouts.

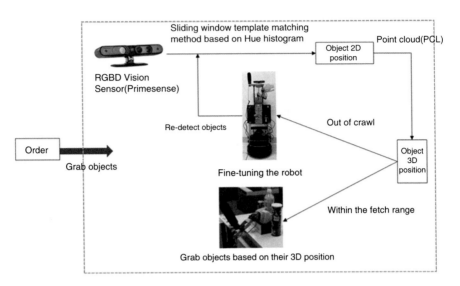

Fig. 12.5 Visual servo grasping of an object

Chapter 13
Integrated Robots Application Case 2: Following and Assisting the User

In addition to being able to accurately identify the user's voice commands indoors, the service robot also needs to accurately identify its user, follow them around, and respond to calls at all times. After following its owner to a relatively unfamiliar environment outdoors, it also needs to have the ability to navigate back indoors to execute tasks autonomously. This places higher demands on the robot's face recognition ability, tracking ability, and navigation ability.

13.1 Case Objectives

In order to adapt to the complexity of the home environment and the complexity of family life, this book designs a task of "following and helping the user to carry goods". The task realizes the robot's function of face recognition, voice recognition, tracking, and autonomous navigation for users. During the test, the robot first uses face recognition to find the user. After finding the user, it prepares to accept commands from the user. After hearing the command from the user to follow, the robot has to follow the user to reach the outdoor area, and after the user commands it to stop, the robot should stop moving and use the robotic arm to grab the items to be carried. Then, the robot takes the items to some room according to the user's command. When the robot reaches its destination, it places the items on the ground and then returns to the user. The task flow of "follow and help user to carry goods" is shown in Fig. 13.1.

© The Author(s), under exclusive license to Springer Nature Singapore Pte Ltd. 2023   279
F. Duan et al., *Intelligent Robot*, https://doi.org/10.1007/978-981-19-8253-8_13

Fig. 13.1 Follow and help
the owner's workflow in
moving the cargo

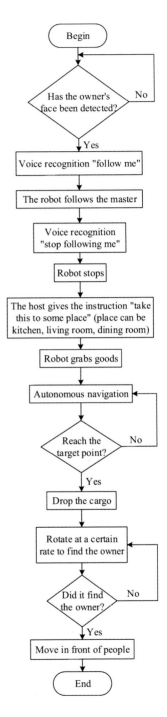

13.2 Voice Recognition Commands

To get a home service robot to obey a user's commands, first build a thesaurus beforehand based on the robot's tasks, such as jack, kamerider, follow-me, stop-following-me, stop-following, stop, kitchen, to-kitchen, bedroom, the bedroom, living-room, the-living-room, dinning-room, to-the-bedroom, to-the-living-room, to-living-room, to-bedroom, take, bring. For how to build a thesaurus and how to perform keyword detection, see Chapter 10. When keywords are detected from voice commands, they correspond to different tasks according to the keywords and send the keywords to the corresponding topics. In this case, "Jack" is the start keyword for speech recognition. When the robot hears the command "Jack, follow me", it needs to extract the keyword "Jack" and "follow-me". The robot will be woken up when it hears "Jack", and the node subscribed to this wake-up topic is started and ready to accept the command; when "follow-me" is detected, the robot issues a message to the topic subscribed to by the following node to start following, and starts follow. Throughout the process, the robot will continue to listen to voice messages. When it reaches the location of the item, the robot receives the command to stop, and the robot stops moving and prepares to pick up the item. After the robot arm picks up the goods, the robot recognizes the keyword about the target location from the voice message and sends the target corresponding to the keyword to the topic corresponding to the navigation node. The robot will continue the navigation process.

13.3 Following and Autonomous Navigation

The implementation of the robot's following function is described in Sect. 7.1 and the principle is the use of a depth camera to acquire a depth image and use a scale controller to control the movement of the robot.

In order for the robot to accurately navigate to the specified location, the working environment is previously mapped and the location coordinates corresponding to each keyword are determined. The environment map and target point settings are shown in Fig. 13.2.

The selection of target points can be achieved by tapping the location on the map and recording the coordinate values returned in the rviz interface, or by moving the robot to the actual target location and recording the coordinates of where the robot is located. After recording the coordinates of all target points, the coordinates of the target points are saved in a text file with the names of the locations corresponding to the target points. In this way, when the topic to which the navigation node is subscribed receives the target point name, it can use the target point name as an index to retrieve the corresponding coordinates in the text file and use this coordinate as the navigation target. The implementation of navigation and map building and other functions is described in Chapter 8.

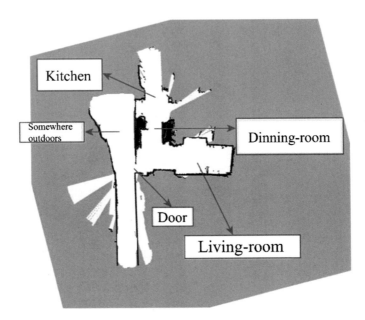

Fig. 13.2 A map that simulates a home environment

In this example, the results show that the robot is good at following, autonomous navigation for obstacle avoidance, and path planning, as shown in Fig. 13.3.

13.4   Detecting and Recognizing Faces

In this example, recognizing the user's face is used as the service robot startup method. The Dlib-based face recognition method (see Sects. 7.4.3 and 7.4.4 for the use of the Dlib package) was used and was implemented well. To detect another user, the OpenCV-based face detection method was used (see Sects. 7.4.1 and 7.4.2 for the use of OpenCV), and the effect was that the robot rotated in place first. If a face is detected, it stops rotating and moves towards the person; if no face is detected, it continues rotating until the robot moves to follow the user. The results of face detection are shown in Fig. 13.4.

a) Tracking users

b) Reach a location outdoors

c) Carry cargo and navigate autonomously   d) Arrive at your destination

Fig. 13.3 Tracking users, autonomous navigation and path planning implementation process. (**a**) Tracking users. (**b**) Reach a location outdoors. (**c**) Carry cargo and navigate autonomously. (**d**) Arrive at your destination

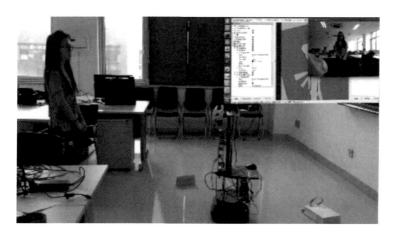

Fig. 13.4 Robot detects a face and moves towards the user

Chapter 14
Integrated Robotics Application Case III: Customers Wave for the Robot Ordering

Service robots work in a wide range of scenarios, and the home environment is just one of them. Service robots working in public face more complex and unfamiliar environments than at home. Robots should be able to respond to commands from strangers and adapt quickly to unfamiliar environments, which requires robots to be able to build maps and complete navigation tasks in unfamiliar environments on-the-fly, with more diverse means of interaction, such as gesture recognition and more sophisticated voice recognition.

14.1 Case Objectives

In this chapter, we design a simulated restaurant environment in which the robot has to perform the task of serving customers' orders in an unfamiliar environment. First, the robot will rotate 360° to build a map of the restaurant and record the current location, i.e., the location of the bar. Two customers are seated at a table and one of them waves to start ordering. The robot has to be able to recognize the wave and move to the waving customer. When the customer wakes up the robot by saying the word "Jack", the robot listens to the customer's menu. Finally, the robot returns to the restaurant bar and repeats the customer's order according to the recorded initial position. The entire workflow is shown in Fig. 14.1.

14.2 Robot Real-Time Mapping

The robot performs real-time mapping during rotation, as shown in Fig. 14.2. The mapping process displayed in rviz is given in the lower left corner of each figure. The robot rotates 360° in place and uses Kinect to initially build the map, obtains

© The Author(s), under exclusive license to Springer Nature Singapore Pte Ltd. 2023
F. Duan et al., *Intelligent Robot*, https://doi.org/10.1007/978-981-19-8253-8_14

Fig. 14.1 Workflow of robot ordering

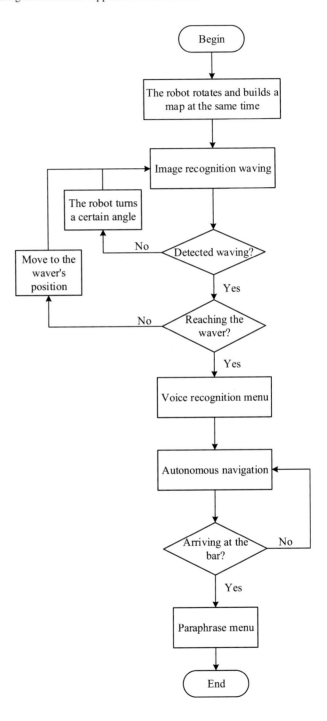

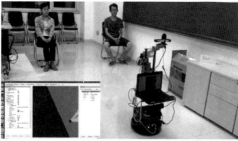

a) Robot initial position and map at this time (lower left corner)

b) Robot rotates 90 degrees and map at this time (lower left corner)

c) Robot rotates 180 degrees and map at this time (lower left corner)

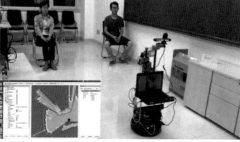

d) Robot rotates 360 degrees and the map is initially established (lower left corner)

Fig. 14.2 The process of real-time mapping for a service robot. (**a**) Robot initial position and map at this time (lower left corner). (**b**) Robot rotates 90° and map at this time (lower left corner). (**c**) Robot rotates 180° and map at this time (lower left corner). (**d**) Robot rotates 360° and the map is initially established (lower left corner)

information about surrounding obstacles, and subsequently records the current location (the bar) in order to later return to the bar to review the menu.

14.3   Robot Recognizes Wavers and Moves to Waving Man

The process by which the robot identifies the waving customer (refer to Sect. 7.2 for an implementation of robot recognition of waving) is shown in Fig. 14.3. A simple proportional controller can be used to control the rotation and advance of the robot based on the angle of offset and the distance, so that the robot gradually moves to the waving person, as shown in Fig. 14.4. In the lower left corner, which is the interface

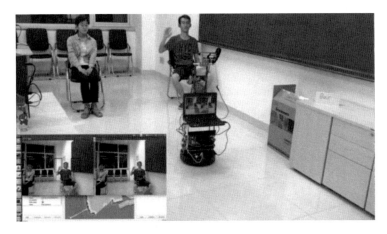

Fig. 14.3 Robot detects two faces but only recognizes one person waving

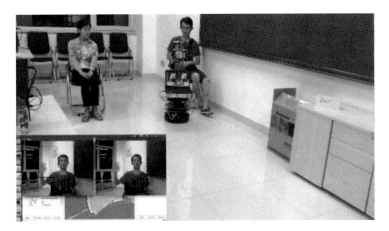

Fig. 14.4 The robot moves to the wielder

between face detection and hand waving recognition, the robot can detect two faces, but only recognizes one person waving.

14.4   Voice Recognition Menu

To identify the dishes ordered by customers, the service robot should first build a thesaurus based on the menu, which should contain wake words as well as words from the menu, such as jack, green-tea, cafe, iced-tea, grape-fruit-juice, strawberry-juice, potato-chips, cookie. The robot extracts the keywords from the customer's speech, matches them with the words in the menu, records the customer's order information and repeats and confirms the menu.

14.5   Autonomous Navigation Back to the Bar

After the customer has successfully ordered by voice, the robot has to return to the initial point, i.e. the location of the bar, based on the recorded initial location and the map builded. When the robot navigates to the bar, it has to repeat the customer's order. The autonomous navigation process is shown in Figs. 14.5 and 14.6.

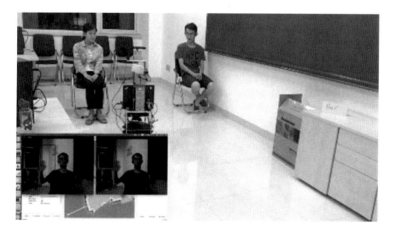

Fig. 14.5 The process of autonomous navigation of a service robot

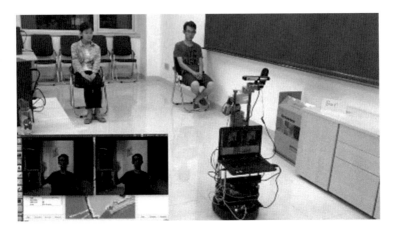

Fig. 14.6 Service robot navigates back to the bar autonomously after ordering

Correction to: Intelligent Robot

Correction to:
F. Duan et al., *Intelligent Robot*,
https://doi.org/10.1007/978-981-19-8253-8

The book was inadvertently published with an incorrect copyright holder as "The Editor(s) (if applicable) and The Author(s), under exclusive license to Springer Nature Singapore Pte Ltd. 2023" whereas it should be "China Machine Press, Beijing, China". The copyright holder has been updated in the book.

The updated version of the book can be found at
https://doi.org/10.1007/978-981-19-8253-8

© The Author(s), under exclusive license to Springer Nature Singapore Pte Ltd. 2023   C1
F. Duan et al., *Intelligent Robot*, https://doi.org/10.1007/978-981-19-8253-8_15

Printed in the United States
by Baker & Taylor Publisher Services